Lucy Hartwell

AGRICULTURE : PEOPLE AND POLICIES

TITLES OF RELATED INTEREST

Agriculture and the European Community
J. Marsh & P. Swanney

The countryside: planning and change
M. Blacksell & A. Gilg

Countryside conservation
B. Green

Development and the landowner
R. Munton

Environmental groups in politics
P. Lowe & J. Goyder

Floods and drainage
E. C. Penning-Rowsell *et al.*

Land policy in planning
N. Lichfield & H. Darin-Drabkin

Landscape meanings and values
E. C. Penning-Rowsell & D. Lowenthal (eds)

London's Green Belt
R. Munton

National parks: conservation or cosmetics?
A. & M. MacEwen

National parks: a greenprint for the countryside
A. & M. MacEwen

Nature's ideological landscape
K. Olwig

Nature's place
W. Adams

Rural housing: competition or choice?
M. Dunn *et al.*

Valued environments
J. Gold & J. Burgess (eds)

AGRICULTURE: PEOPLE AND POLICIES

Edited by

Graham Cox, Philip Lowe
and Michael Winter

London
ALLEN & UNWIN
Boston   Sydney

This book was typeset, proofed and passed for press by the editors.

**Allen & Unwin (Publishers) Ltd,
40 Museum Street, London WC1A 1LU, UK**

Allen & Unwin (Publishers) Ltd,
Park Lane, Hemel Hempstead, Herts HP2 4TE, UK

Allen & Unwin Inc.,
8 Winchester Place, Winchester, Mass. 01890, USA

Allen & Unwin (Australia) Ltd,
8 Napier Street, North Sydney, NSW 2060, Australia

First published in 1986

British Library Cataloguing in Publication Data

Agriculture: people and policies.
1. Agriculture and state−Great Britain
I. Cox, Graham II. Lowe, Philip
III. Winter, Michael
338.1′841   HD1927
ISBN 0−04−307001−9

Library of Congress Cataloging-in-Publication Data

Agriculture: people and policies.
Includes bibliographies and index.
1. Agriculture and state−Great Britain−Congresses.
I. Cox, Graham II. Lowe, Philip. III. Winter, Michael.
HD1927.A345 1986   338.1′841   86−20634
ISBN 0−04−307001−9 (alk. paper)

Printed in Great Britain by
Anchor Brendon, Tiptree, Essex

Acknowledgements

We gratefully acknowledge the support of the Rural Economy and Society Study Group of the British Sociological Association and the Economic and Social Research Council in the preparation of this volume. Our thanks also go to Trudy Mellor for her painstaking care in preparing the typescript.

Contents

1 The state and the farmer: perspectives on agricultural policy

GRAHAM COX, PHILIP LOWE and MICHAEL WINTER

INTRODUCTION

Relations between the state and the farmer are currently the stuff of news headlines. There is an emerging public consciousness that, apart from the nationalised industries, agriculture is the most highly regulated and state supported sector of the economy. Concern has been heightened with the realisation that government involvement has had considerable implications for the organisation and structure of farming, for the welfare of rural communities, for the ecology of the countryside, for the price and availability of food, for nutritional standards, and for international trading relationships.

Previously, some of these issues were less salient. In any case the relationship between the farmer and the state was for the most part rendered opaque by the technical complexity and intricacy of the policy instruments employed. In addition, the farmer's own self-image as the independent yeoman wresting a living from the soil was mirrored in popular and artistic perceptions. In sharp contrast to this image of rugged individualism, it is now evident that the farmer is caught in a web of relationships which extends, in one direction to Whitehall and Brussels, in another to the big chemical manufacturers and food conglomerates, in another to the banks and credit companies, and in another to the flora and fauna of natural ecosystems.

The purpose of this volume is to explore some of these relationships. For a variety of reasons, mainstream social science

has tended to ignore such issues as part of a more general neglect
of rural problems. To overcome this intellectual lacuna, the Rural
Economy and Society Study Group (RESSG) was established in 1979 to
provide a forum for all those studying the social formation of rural
areas in advanced societies. It comprises social scientists from a
variety of disciplines including sociology, geography, economics,
politics, anthropology and planning. Formed amidst a renaissance of
rural social research, the group has sought to encourage
theoretically informed analysis of rural issues, and to foster a
more integrated and critical approach than has traditionally
characterised rural studies. On the subject of agriculture, this has
inevitably involved confronting some of the dominant approaches and
assumptions of agricultural economics, and below we summarise the
key points in this intellectual challenge. Though our comments may
seem harsh, they are presented in the spirit of stimulating
constructive debate. We are therefore pleased that the book includes
important contributions from agricultural economists (as well as
from other social scientists) which demonstrate some of the
strengths and robustness of that discipline.

Most of the papers in the book were presented at a conference
convened by the Rural Economy and Society Study Group under the
title 'Agricultural Organisation in the Advanced Societies'. It was
held in Oxford in January 1985 with the support of the Economic and
Social Research Council. How is agricultural production organised,
socially and economically? How is it affected by government
policies? And what are the determinants of policy towards
agriculture? These were the key questions participants addressed. In
selecting papers for the book and asking authors to revise their
original contributions, the aim has been to produce a wide-ranging,
yet rounded and integrated, volume representative of the breadth and
vitality of current research. Thus, the book should provide a useful
point of reference for the burgeoning interest in agricultural
policy. Our intention in this introduction is to relate the
collection as a whole to past research and to comment on its
implications for the direction of future research.

THE LEGACY OF AGRICULTURAL ECONOMICS

The design and assessment of agricultural policy measures has
traditionally been the concern of agricultural economics. Indeed,
the economic instruments of policy are one of the central foci of
their discipline. In addition, it is almost mandatory for text-books
in agricultural economics to present a descriptive review of the
policies and legislation affecting farming, and generations of
students have been schooled on such material. Increasingly the
analysis has been within an econometric framework. While this has
provided some valuable insights into specific policies, all too
often it has yielded predictive models which either ignore, or make
limiting assumptions about, the context and determinants of policy.
Indeed, the discipline has shown a distinct aversion towards
considering the politics of agriculture, and there has been no
development of a general political economy of the industry.

One cause of the technocratic bent of agricultural economics is
the role that the discipline has come to play in the policy process
itself. In the words of one commentator, 'the massive advance of
state economic intervention required the systematic provision of
quantitative information in a form suited to guide the determination
of economic policy...econometrics may be said to have come of age'
(Grahl 1979). This is not to suggest that agricultural economics has
slavishly provided government with statistical 'fodder' to justify
its policies. On the contrary economists have often despaired at the
cavalier way in which their analysis is treated by politicians,
although relations are generally more cordial with the permanent
officials of the Ministry of Agriculture, Fisheries and Food (MAFF).

Until recently, university departments of agricultural
economics (apart from the Oxford Institute) were financed directly
by MAFF. Even now much research is funded by the Ministry and
several universities still supply it regularly with farm management
survey data. This has given research a narrowly 'practical'
orientation, and, in the past, often resulted in senior Ministry
officials being appointed to chairs. One consequence has been that
some agricultural economists have too readily accepted MAFF's
assumption that the prime policy objective should be to raise the

cost-efficiency of agriculture, regardless of the social and
environmental implications. Close links with the Ministry have not
encouraged agricultural economists to undertake the sort of
macro-economic or institutional analysis which might cast a critical
light on such basic policy parameters (for a discussion of parallel
processes in the USA, see Newby 1980, 1982, 1983).

A recent House of Lords Select Committee (1984) on
'Agricultural and Environmental Research' was highly critical of
'the closed loop between Agriculture departments and agricultural
research' whereby the departments' concerns so dictated research
priorities. This may have suited the Ministry during the period when
the objectives of agricultural policy seemed straightforward and
uncontentious. Now, however, it finds itself bereft of the
background research and analysis that it needs in order to respond
constructively to the new and diverse pressures on agricultural
policy, pressures to curb overproduction, to promote rural welfare,
to conserve the environment and to improve nutritional standards.

It is no coincidence that the recent path-breaking critiques
which have succeeded in shifting the agenda of agricultural policy
have come from outside agricultural economics. Marion Shoard, author
of 'The Theft of the Countryside' (1980), is from a planning
background. Howard Newby, author of 'Green and Pleasant Land'
(1979), is a prominent sociologist. Sir Richard Body, a somewhat
maverick backbench Tory MP, has contributed 'Agriculture: The
Triumph and the Shame' (1980) and 'Farming in the Clouds' (1982).
John Bowers and Paul Cheshire, whose 'Agriculture, the Countryside
and Land Use: An Economic Critique' is the most sophisticated of the
critiques, are both mainstream economists. They make that point
quite clear in the rather barbed disclaimer in the preface to the
book to the effect that 'neither author has any connection with
departments of agriculture or agricultural economics in our
respective universities'. One exception to this pattern is provided
by the work of the agricultural economist, Richard Howarth, whose
book 'Farming for Farmers' (1985) is a thoroughly hard-hitting
attack from a monetarist position.

To be fair also, all these authors draw to a greater or lesser

extent on solid technical work from within agricultural economics.
Indeed, the subject represents an invaluable source of analysis and
data on the (often arcane) operations of agricultural policy and the
workings of the farming economy, a source which those who wish to
develop a sophisticated and integrated analysis of agricultural
policy cannot afford to ignore. In the main it covers what might be
termed the middle ground (i.e. the detailed content and workings of
policy measures) but omits the start and the finish (i.e. the
context and process of policy formation and the wider consequences
and ramifications of policy). The sort of integrated analysis needed
to encompass all of these elements requires an interdisciplinary
approach aware of the insights furnished by sociology, economics and
political science.

TOWARDS A POLITICAL ECONOMY OF AGRICULTURE

To achieve such a synthesis it will be necessary to bring into
critical focus some of the (often implicit) assumptions of
agricultural economics concerning the ways in which policy is formed
and the motivations of agricultural producers. The most basic of
these assumptions derive from neo-classical economics[1] and posit
narrowly instrumental behaviour on the part of individual producers
and a facilitating role for the state of ironing out market
imperfections (Nou 1967, Newby 1982). An important stream in rural
sociology has been taken up with qualifying the former assumption:
investigating, for example, the apparent 'irrationality' of farmers
and their 'maladjustment' in the face of an agricultural policy that
takes for granted their overriding economic rationality as profit
maximisers. A great deal of the sociological study of family
farmers, small farmers and part-time farmers has been in this vein
(Gasson 1971).

Other efforts to elucidate what agricultural economists have
euphemistically dubbed 'human factors' or 'residual social factors'
in agricultural production have highlighted the diversity and
heterogeneity of the farming industry. This is recognised in
agricultural economics in the concern over such issues as the
relationship between farm size and efficiency. But there is a
neglect of qualitative differentiations between enterprises based on

the social relations of production, rent relations, links with other
forms of capital, and so forth. For example, recent sociological and
anthropological work on the sexual division of labour within the
farm family and on the development cycle of the domestic group have
considerable implications for our understanding of the production
process, emphasising inter alia the temporal fluctuations in the
demand for and availability of family labour (Bouquet 1985,
Friedmann 1978, 1980, Marsden 1984). Yet many agricultural
economists continue to take the fixed labour input of family farms
over time as a reasonable assumption. Other agricultural economists,
it should be said, have made substantial empirical contributions to
agrarian sociology, though they have often been on the fringes of
their discipline, most notably within farm management.

The other assumption that needs to be subjected to critical
scrutiny is that the rationale for state intervention is as a
corrective, to overcome any structural obstacles to perfect
competition. The orthodox account, rather than analysing the social
or political context of intervention, begins with a depiction of how
actual conditions fall short of those necessary for market
equilibrium, and supposes this to be the premise for government
intervention. As such it presents an ex post facto justification for
government action rather than an analysis of its concrete causes. In
specific instances the failure of market competition may be one of
the reasons for state intervention, but this cannot furnish a
general explanation of state involvement, for two basic reasons.
First it does not measure up to the comprehensive nature of
government intervention in the agricultural sector. Secondly it
overlooks the long history of state involvement, predating the
development of competitive, capitalist agriculture (Hopkins and
Puclhala 1978).

Any analysis of policy, therefore, must recognise the state as
a primary and active (not merely a reactive) agent in the
agricultural sector. From this viewpoint and in comparative
perspective, Victorian Britain appears as an aberration, its
laissez-faire and free trade approach towards agriculture, though
not towards land, dependent upon a unique conjuncture of political
and economic forces. It is perhaps unfortunate therefore that the

grand theories of agriculture, whether from a classical, liberal or Marxist perspective, are rooted in the perception that nineteenth century British farming represented the apotheosis of modern agriculture.

These observations point to the need for a political economy of agriculture, by which we mean a methodological perspective which recognises the dialectical interaction between political and economic systems. Thus, markets are structured by the political decisions and actions of various agencies of the state. They, in turn, are influenced and constrained by external groups whose power is related to their structural position in the economy as well as to the strength of their political organisation.

In considering how to integrate these various analytical elements the heuristic model put forward by Newby (1982) of the role of the state in the changing structure of agriculture is instructive (Fig 1). The most notable feature of the model is that it both starts and finishes with the social structure of agriculture, and is thus dynamic and cyclical rather than linear. It also recognises that agricultural producers are neither passive recipients of policy nor its sole determinants. The social structure of agriculture, it is postulated, gives rise to a certain array of organised or 'institutionalised' interest groups which are among a number of influences on state policy formation. In its turn policy differentially affects the conditions of production of different groups of food producers, so contributing to changing social relations and social structures. The precise balance between agriculture's and the state's determination of policy outcomes is an empirical question and can best be understood through the analysis of the 'survival strategies' of producers. These form 'the analytical link between causes of changing structures, the nature of state intervention in organising agricultural production, and the changing class structure of the rural sector'. These ideas are predicated on a particular non-instrumentalist view of the state. The state is not necessarily seen as serving a particular class, nor as a unity in itself. The chapters which follow illuminate various facets of this model.

Figure 1 Changing Structure of Agriculture in Advanced Industrial Societies (2)

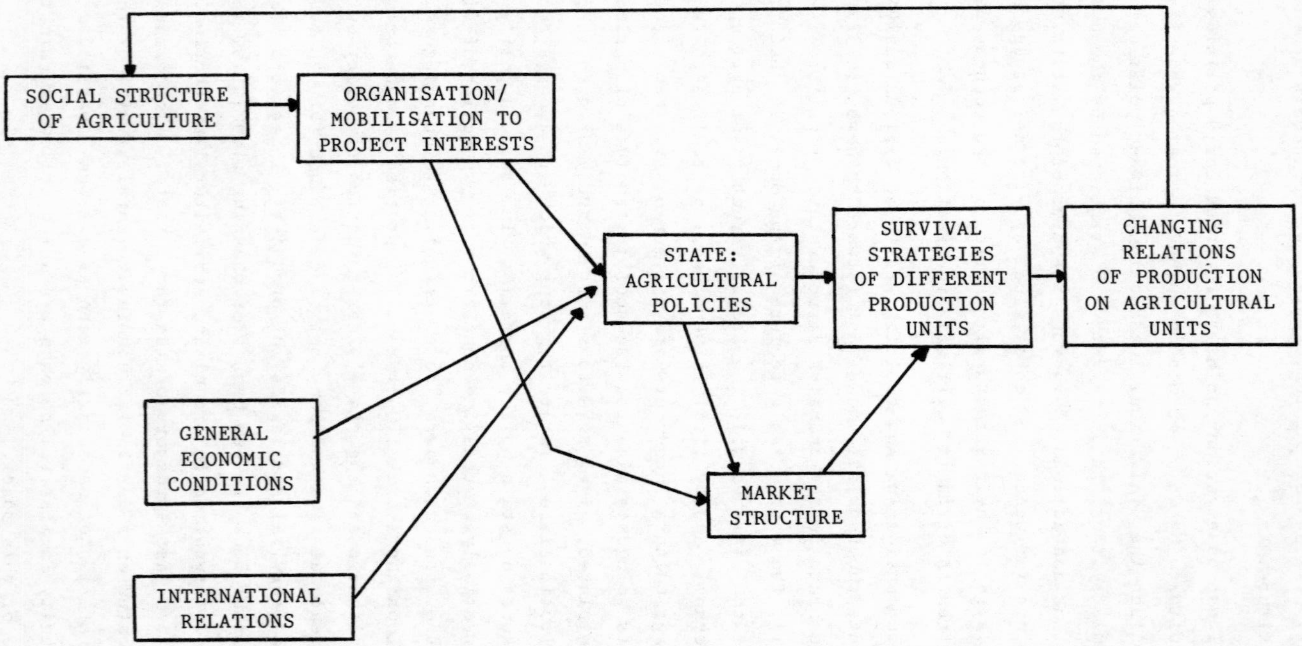

THE CONTRIBUTIONS

Issues relating to the structure of farming are the first concern of the volume. In recent years the organisation of agriculture under advanced capitalism has emerged as a major research focus, stimulated not least by the structural changes that have occurred in Western agricultures. In particular, the relentless decline in the number of farms has attracted attention to the nature and prospects of a sector of the economy where small, family producers still predominate. A key research problem, therefore, has been to determine the special resilience of the family farm and the factors undermining its viability.

In thinking about the organisation of farming and its economic relationships in the advanced societies, a number of authors have found it fruitful to adapt concepts and theories from development studies, particularly ideas about peasant society and the peasant economy. Some of the more prominent in this endeavour are represented here. Harriet Friedmann, in her previous writings (1978, 1980, 1986a), has expounded the concept of 'simple commodity production' to denote market-oriented producers reliant on family labour. On the one hand, this theoretical category is distinguished from capitalist production by the absence of systematic recourse to wage labour. On the other hand, it is distinguished from peasant or subsistence production by the production of commodities for the market and dependence on exchange relations for the inputs to production, as well as for the requirements of personal consumption. Thus, for simple commodity production to exist, labour, land, input and output markets all have to be fully developed. It therefore depends for its reproduction on the conditions set by the wider capitalist mode of production, as well as on the internal goals and social relations of production within the farm household even though (in the absence of wage labour) it does not itself embody capitalist relations of production.

In the first contribution to the volume (Chapter 2), David Goodman and Michael Redclift analyse the recent development of Marxist theories about agriculture, paying particular attention to Friedmann's concept of simple commodity production. Though they

praise its clarity, they challenge its status as a distinct
theoretical or empirical category, arguing instead that it is a
historically contingent expression of the relationship between
technology and the rural labour process. Specifically, Goodman and
Redclift reject Friedmann's argument that simple commodity
production has no internal class relations, charging that she
underplays the importance which wage labour can assume in the
reproduction of the family farm for extended periods of the
household demographic cycle, when family labour is inadequate.
Goodman and Redclift seek to shift the focus of debate from the
internal structure of the family enterprise, to the determinants of
the changing technical basis of production. Because of the objective
limits that organic nature and land place on its capacity to
transform the production process, capital has been forced to pursue
partial industrial appropriations of the rural labour process which
are then reincorporated in agriculture as valorised inputs or
produced means of production. In this way, capital has progressively
appropriated activities and value once regarded as essential to
farming. As constraints to subsumption have been overcome, so
farming has been increasingly penetrated by capital and its scope
has been progressively refined. These processes, Goodman and
Redclift suggest, have been obscured by a powerful ideology
regarding the survival of the (supposedly unchanging) family farm.

Elsewhere, Friedmann (1986b) has defended the project to
construct a concept of simple commodity production against these
criticisms. She accepts Goodman and Redclift's emphasis on organic
nature, land and space and the tendential subordination of these to
capital, but argues that this perspective and simple commodity
production are not alternatives but complementary dimensions of a
more complete analysis, the latter for the analysis of family
enterprises generally and the former for their specific place in
agriculture. Her paper in this volume (Chapter 3) addresses the
question of what is specific to family enterprises in general in
capitalist agriculture, and identifies two distinctive features:
first, production is organised through kinship; and second, property
and labour are combined. The dynamics of family enterprises stem
from the interplay of household and business. Thus, at all times,

but especially across generations, the enterprise is dependent upon family formation and development for its survival. In turn, the unity of kin group and work group creates at least dual roles for each member and patterns a division of labour and inequalities of control and ownership, based on gender and age relations. What is not distinctive about family enterprises, however, is their dependence on the labour market for any additional labour needs. But, Friedmann insists, as long as property and labour are combined, through the owner continuing to work in production, a transition has not occurred, and she reviews the strategies pursued by family farms to reduce their dependence on the labour market. Finally, the contradictory unity of property and labour creates contradictory possibilities for political alliances, and Friedmann discusses the implications for such issues as women's equality, agricultural labour, ecology, food politics, and the ideology of private property.

In the next contribution (Chapter 4), Michael Winter examines the historical background to specialised, market-oriented, family farming in Britain, focusing on developments in West Devon. In this area of pastoral farming, landlordism and capitalist relations of production predominated during the nineteenth century, but there was also a sizeable proportion of small farmers, including significant numbers of owner-occupiers, who were reliant on family labour and had a quasi-subsistence orientation. The Great Depression of the late nineteenth century had a major impact on this pattern, undermining the economic position of the large landowners and favouring small-scale producers. The consequent emergence of simple commodity production, Winter suggests, represented the convergence of two separate groups pursuing different survival strategies: on the one hand, small farmers and owners adopting an increased market orientation, partly in response to new market opportunities; and on the other hand, medium-sized capitalist farmers shedding hired labour and cutting costs through mechanisation, in response to deteriorating product prices.

The survival strategies of contemporary farmers is a common focus of the next two contributors. With agricultural incomes being squeezed and facing the possibility of relative decline in the medium term, then increasingly part-time farming has become the subject of intense discussion for what it offers both as a survival or income optimising strategy for farm families, and as a policy option for maintaining rural land use and rural populations. As Ruth Gasson (in Chapter 5) easily shows, however, the meanings of the term 'part-time farming' in current usage are as various and often misleading as the stereotypes with which it is associated. Defining part-time farming as families running the farm as their own business with one or more members having another gainful activity, she presents findings from a study of 345 part-time farming households visited during the summer of 1981. As against the popular misconceptions it is clear that part-time farming can be a stable and permanent status as well as a transitional phase and, moreover, one that is positively sought rather than merely tolerated. Contrary to another popular notion, part-time farming does not, typically, represent the proletarianisation of the family farmer. Britain may be distinctive in this respect, however, since the proportion of farmers with non-farm jobs who were self-employed is much higher at 45 per cent than the 26 per cent in the EEC as a whole. An analysis of the dynamics of part-time farming enables further myths to be challenged, and the typical part-time farmer emerges as someone who runs another business, most likely off the farm and not necessarily linked in any way with agriculture. There is considerable diversity though with part-time farming more likely to be prevalent on both very small and very large farms than on medium sized family farms of between 50 and 500 acres. Gasson concludes by looking at the wider social, environmental and resource use implications of the fact that part-time farming involves about a quarter of the nation's farming families and farm businesses. The tendencies identified in the paper suggest that part-time farming may be associated with greater stability for rural employment and population, and a slower rate of change in the appearance of the countryside.

Many of the generalisations about U.K. agriculture overlook or gloss over Northern Ireland, even though as Joan Moss points out (in Chapter 6) farming contributes much more to economic activity there than in Great Britain. Indeed, its farming structure is more akin to those of other European countries, not least that of neighbouring

Eire (due, of course, to their common history and the same late nineteenth century tenurial reforms which created a class of peasant proprietors). 'Small scale' farms averaging only 17 ha of crops and grass account for almost half the province's farm businesses, a third of its full-time farm labour (as reported by farmers in the agricultural census) and a seventh of its total standard gross margin. The very small incomes that these farms generate in most cases necessitates off-farm employment and therefore beef cattle and sheep production predominate owing to their relatively low and less routinised labour requirements. Moss addresses the role and viability of these small-scale farmers drawing on a survey of 200 of them. Some of the findings parallel those of Gasson (see Chapter 5), especially regarding the stability of part-time farming and the diversity of types of off-farm employment. But a more fraught picture emerges from the cold statistics: of the difficulties of finding additional employment in the remoter hill areas and in a region of very high unemployment; of a third of the small farm households being dependent on such transfer payments as old age pensions and family income supplements; of the postponement of marriage and family formation; and of the necessity, for many of those with families, to work long hours both on and off the farm to achieve an acceptable standard of living. Even so, these small-scale farms offer a certain security (as well as being a capital asset) which ensures their persistence. The chapter concludes with a consideration of measures which might assist small scale farmers, emphasising that if policy makers consider it desirable to maintain rural populations, then policies which currently discriminate against the smaller farmer will have to be reassessed.

From the organisation of the labour process, we turn in the next two contributions to the position of land and landownership relations in agricultural production. In Chapter 7 Sarah Whatmore focuses on the major manifestations of the rent relation in contemporary British agriculture. She adopts a Marxist analysis seeing the rent relation as part of the contradiction between private property and capital; the development of capitalist agriculture being propelled by the need to circumvent this contradiction, through reducing the dependence of production and accumulation on land. Periodic restructurings of landownership relations have in the past overcome the immediate barrier to investment presented by established ownership structures, but the basic contradiction of the rent relation has remained though manifested in a different form. Thus, the growth of owner-occupation, combining the roles of capitalist and landowner, provided a solution to the charged conflict between high rents and falling incomes that beset the traditional landlord-tenant system at the turn of the century. Yet the farmer who invests in land as an essential means of production still experiences the contradiction of the rent relation. The purchase of land, at prices determined not simply by agricultural use value but by the status of land rights as fictitious capital, detracts from the capital available for productive investment, leading to liquidity crises. The standard response of farmers who use land as collateral to secure a bank loan or a mortgage, establishes a credit relation equivalent to a rent relation. The recent advent of institutional landownership, reviving the landlord-tenant relationship, has been seen as one way of relieving the liquidity crisis and overcoming the problems that farmers face with high land values. However, as Whatmore points out, the rationale behind institutional landownership lies in the particular development of finance capital in Britain in the post-war period. The financial institutions have been highly selective in the pattern and form of their investments. Moreover, their strictly financial interest in agricultural land and its marginal position within their total asset structure means that they are the most volatile influence in the agricultural land market.

A significant, if declining, feature of British agriculture is the private landlord-tenant system. In Chapter 8, Terry Marsden examines the political and economic forces that have shaped the legislation governing this form of tenure. He sets his analysis in a general discussion of the dynamics of property-state relations. Periodic state intervention to restructure and legitimise private property rights is required to mediate the contradiction between democratic principles and capitalist processes which necessitate the commodification of land rights and their concentration in the hands of a minority. At times this has entailed restricting property rights; at other times, aiding the process of concentration. Overall, the primacy of private ownership and control of land has been maintained. Marsden shows how these processes have been evident in state action in modifying landlord-tenant relations. The analysis covers developments in Britain since the turn of the century, but focuses specifically on the pressures which led to the removal in 1984 of statutory successional rights for tenants that had been introduced eight years earlier. Marsden argues that this was the consequence of the successful mobilisation by landowning interests of ideological arguments which played down notions of family continuity and stressed instead the need to revitalise the farming ladder. However, he casts doubt on whether the change will encourage the creation of new tenancies, and suggests that its real import lies in the restoration of landlords' control over their land.

The structure of farming and of landownership establish the social context in which agriculture is practised, and the next contribution (Chapter 9) examines factors influencing farmers' decisions regarding the management and development of their farms. Clive Potter's concern is to capture analytically the dynamics of change in the countryside. Investment and management decisions are the outcome of a complex and continuous interaction between agents and structures and satisfactory understanding needs to embody insights from a number of levels of explanation which have, hitherto, typically been examined separately. Based on a study of 175 farmers in three lowland counties he examines the countervailing impacts of investment in land improvement and conservation advice and shows that it is possible to identify different 'investment styles', themselves the outcome of complex sets of policy influences, and opportunities and constraints at the level of the individual farm enterprise. When assessing the range and quality of conservation advice Potter focuses on the role of the Farming and Wildlife Advisory Group comparing farmers who are involved, or who have received advice, and those so far unaware of its philosophy. Finally, he considers the policy implications of his findings emphasising that the challenge to policy makers is to recognise the complex nature of influences on farm management and to introduce a system of controls, state aids and subsidies which can simultaneously promote farm income, conservation and wider social objectives.

The social and environmental consequences of agricultural policy are currently matters of great contention, and the next two chapters examine some of the issues and conflicts involved. In a self-consciously polemical riposte to those who have attacked the profligacy and damaging environmental implications of agricultural policy and have presented farmers as enemies of the general interest George Peters (Chapter 10) outlines production, income and price trends in the post-war period. He then examines, in turn, the current position of the farmer and the consumer and uses his conclusions as a basis both for placing the debate about agriculture and the countryside in a broader perspective and casting doubt on the claim that a reduction in support for agriculture would, of itself, be beneficial in conservation terms.

The political dynamics of the conflict between agriculture and rural conservation is the subject of our own contribution to the volume (Chapter 11). In an analysis of the opposing political forces, we present successive accounts of the relevant policy communities and the developing stances they, and their constituent members, have adopted. The policy community for rural conservation is characterised as large, diverse and pluralistic; that for agriculture, as small, tightly-knit and corporatist. Environmental concern over the consequences of contemporary agricultural policy and modern farming practices has increasingly brought these two communities into contact. The paper analyses the different tactics pursued by the constituent groups, as conservationists have sought some influence in agricultural policy making and a more detailed legislative framework for land use management; and as agricultural interests have sought to defend the autonomy of the farmer in making production and land use decisions, and the autonomy of the Ministry of Agriculture and the farming community in the administration of agricultural policy.

One dimension of the recent politicisation of agriculture is a growing polarisation between the political parties on the subject. In Chapter 12, Andrew Flynn reviews the development of agricultural policy and party politics in Britain in the post-war period. From extensive archival research, he challenges the view that party politics have had marginal impact on agricultural policy making and that little has separated the two main parties in their farming policies. Instead, he reveals complex manoeuvrings both between and within the political parties as they have striven to appeal for rural votes and to satisfy their supporters and activists. Flynn sets his discussion in the context of theories of party competition. However, he shows that none of these theories is adequate to explain his findings on policy generation within the Labour and Conservative parties: party policy making is more complex than these theories would allow. The extra factors he would adduce, besides the competitive nature of party politics, include the internal structure of the parties, their ideological commitments, and the influence of interest groups.

Agriculture currently has a prominent political profile and the parties present competing visions of the future for farming and for rural areas. Thus, whereas the Conservatives look to greater economic efficiency in agricultural support and the diversification of farm enterprises, Labour would reduce support in favour of consumers and rural job creation and would introduce planning controls over agriculture and the Liberal Party would redirect the available funds towards the small farmer and conservation work. This, at the very least, offers a democratic choice, but it augurs major changes in the organisation and operation of farming.

The next few years are likely to be critical for all those involved in agriculture - men and women in production, policy makers and the public in general. It is vital that good policy related research is undertaken. It is to be hoped that this volume provides some examples of the high quality of research already emerging.

FOOTNOTES

1   This is not the place to enter into a lengthy exposition on the
    nature of neo-classical economics. For a critical discussion see
    Hollis and Nell 1975. We are not suggesting that neo-classical
    economics is a single unifying theory - rather it is a broadly
    based tradition which offers a set of methodological tools
    capable of use in a variety of ways. The assumptions we describe
    here are attributable to the particular expression which
    neo-classical economics has taken in British agricultural
    economics. For some very different, and highly stimulating,
    strands of neo-classical treatments of agriculture see Nou 1967,
    Donham 1981, Durrenberger 1984.

2   We are grateful to Howard Newby for permission to use this
    diagram (see Newby 1982).

REFERENCES

Body, R. 1982. Agriculture: the triumph and the shame. London:
    Maurice Temple Smith.
Body, R. 1984. Farming in the clouds. London: Maurice Temple Smith.
Bouquet, M. 1985. Family, servants and visitors: the farm household
    in nineteenth and twentieth century Devon. Norwich: Geo Books.
Bowers, J. K. and Cheshire, P. 1983. Agriculture, the countryside
    and land use. London: Methuen.
Donham, D. L. 1981. Beyond the domestic mode of production. Man 16,
    515-541.
Durrenberger, E. P. (ed) 1984. Chayanov, peasants and economic
    anthropology. Orlando Florida: Academic Press.
Friedmann, H. 1978. Simple commodity production and wage labour in
    the American plains. J. Peas. Stud. 6, 71-100.
Friedmann, H. 1980. Household production and the national economy:
    concepts for the analysis of agrarian formations. J. Peas.
    Stud. 7, 158-184.
Friedmann, H. 1986a. The family farm in advanced capitalism: outline
    of a theory of simple commodity production in agriculture. In
    The political economy of agriculture in advanced industrial
    societies, F. H. Buttel and T. Murphy (eds). New York:
    University Press of America.
Friedmann, H. forthcoming 1986b. Patriarchy and property: a reply to
    Goodman and Redclift. Sociologia Ruralis 26.
Gasson, R. 1971. Use of sociology in agricultural economics.
    J. Agric. Econ. 22, 28-38.
Grahl, J. 1979. Econometric methods in macroeconomics: a critical
    appraisal. In Issues in political economy, F. Green and P. Nore
    (eds), 3-32. London: Macmillan.
Hollis, M. and Nell, E. (eds) 1975. Rational economic man: a
    philosophical critique of neo-classical economics. Cambridge:
    Cambridge University Press.
Hopkins, R. F. and Puchala, D. J. (eds) 1978. The global political
    economy of food. Madison, W I: University of Wisconsin Press.
House of Lords 1984. Agriculture and the environment. 20th Report.
    Session 1983-84. Select Committee on the European Communities.
    London: HMSO.

Howarth, R. 1985. Farming for farmers? London: Institute of Economic Affairs.

Marsden, T. 1984. Capitalist farming and the farm family: a case study. Sociology 18, 205-224.

Newby, H. 1979. Green and pleasant land? Social change in rural England. London: Hutchinson.

Newby, H. 1980. Rural sociology - a trend report. Current Sociology 28, 1-141.

Newby, H. 1982. Rural sociology and its relevance to the agricultural economist, a review. J. Agric. Econ. 33, 125-165.

Newby, H. 1983. The sociology of agriculture: towards a new rural sociology. Annual. Rev. Sociol. 9, 67-81.

Nou, J. 1967. The development of agricultural economics in Europe. Uppsala.

Shoard, M. 1980. The theft of the countryside. London: Maurice Temple Smith.

2 Capitalism, petty commodity production and the farm enterprise

DAVID GOODMAN and MICHAEL REDCLIFT

INTRODUCTION

Marx and Engels wrote very little about agriculture. If we exclude
the very extensive discussion of ground rent in the third volume of
Capital, we are left with a couple of fragments in the first volume
of Capital and three polemical, although penetrating, essays - The
Eighteenth Brumaire, The Peasant Question in France and Germany and
the Critique of Gotha Programme. Perhaps it is not surprising, then,
that Marxists have spent so much time interpreting, and elaborating
upon, what Marx and Engels wrote. Lenin, Kautsky, Preobrazhensky,
Kritsman, Rosa Luxemburg, Gramsci, Mariategui ... the list is a long
one.

This paper analyses the recent development of Marxist theory
towards agriculture in general, and simple commodity production in
particular. It reviews the problems which historical cases have
presented for Marxist theory and examines the consequences of these
theoretical refinements for an understanding of contemporary
agrarian change in developed economies. Finally, we examine some
recent attempts to define the structural position of the capitalised
family farm under capitalism, and attempt to theorise, in a more
convincing fashion, the relationship between technology and the
rural labour process in the Western industrialised countries. Even a
superficial reading of the literature suggests that the development
of agriculture under capitalism presents problems for Marxist theory
(Goodman and Redclift 1981). As an initial step towards elucidating

these problems it is helpful to review the substantive claims of
orthodox Marxism, as well as the invitations to revisionism in the
'classic' literature.

MARXIST THEORY OF AGRICULTURE: STRENGTHS AND WEAKNESSES

In the Eighteenth Brumaire of Louis Napoleon Marx set out the
reasons for believing that the French peasantry would be destroyed.
This was in 1852. Some sixteen years later he set out his reasons
for believing that large-scale capitalist agriculture was superior
to small-scale petty commodity production, in the first volume
(1868) of Capital. Almost thirty years after this Engels returned to
these issues in his essay, The Peasant Question in France and
Germany, which was written in response to the efforts European
Socialist parties were making to draw political support from peasant
farmers. The debates within German social democracy in the 1890s and
the later discussions in the Soviet Union between the 'Agrarian
Marxists' and supporters of the orthodox Leninist school of thought,
not only called up Marx and Engels as witnesses, but sought to
elaborate upon some key issues in Marxist theory. The key problems
were:

(a) If large-scale production was superior to family production
    in agriculture, why had the family producer not
    disappeared?

(b) Could family production relations continue to co-exist with
    large-scale agricultural production under capitalism?

(c) What tendencies were evident in the differentiation process
    in agriculture, first discussed by Marx, but more fully
    developed by Lenin in The Development of Capitalism in
    Russia at the turn of the century?

As we shall see each of these questions is of considerable
importance in the contemporary debate about the future of 'family
farming' and the theoretical status of petty commodity production in
the agricultural development process.

According to Marx, capitalist production drives a wedge between
agriculture and industry. The industrial proletariat is located in
towns, but 'the dispersion of the rural labourers over large areas
breaks their power of resistance' (Marx 1970:506). The political

struggle, then, is largely confined to the urban proletariat. Nevertheless 'the desire for social changes and class antagonisms are brought to the same level in the country as in the towns' (ibid). Eventually industry and agriculture are recombined under capitalism, on the basis of the characteristics they acquired through separation. Agriculture, far from being an exception to the rule of capitalist penetration, provides a particularly good example of it.

A later passage in Capital Volume I, elaborates on the way in which agriculture is transformed. Marx writes that 'centralisation completes the work of accumulation by enabling industrial capitalists to extend the scale of their operations' (Marx 1970:627) and suggests that similar tendencies can be observed in agriculture, providing examples from Ireland where, in the aftermath of the Irish Potato Famine, farms under 15 acres disappeared (ibid:711). Again, agriculture obeyed the same laws as industry. Ultimately they would become indistinguishable. Marx is drawing attention to the characteristics of the production process in agriculture which favour large-scale enterprises and expanded reproduction, through reinforcing the same tendencies towards concentration and centralisation that were observed in industry. His discussion of ground rent in Volume III makes it very clear that different farms have varying potential as capitalist enterprises. Nevertheless there is nothing, either in Volume III or Volume I, to suggest that agriculture would not be organised along capitalist lines in the fullness of time.

This observation is true not only of Marx's wider discussion of agriculture and industry, but also in what he says about the peasantry. The peasantry, discussed with such vigour in The Eighteenth Brumaire, was the principal target for rural capital. If production capital was part of the inexorable logic, so too was merchant capital, which reduced the peasantry to a vehicle for accumulation. Changes in smallholding property, which served to emphasise market relations rather than subsistence, served to alter the relations between the French peasantry and other classes in society (Marx 1968:173). The French peasantry had ceased to be independent under Louis Napoleon and had become the means by which

merchant capital gained in strength, while the state gained a
political constituency. Merchant and finance capital, as well as
industrial capital, would play an important role in the eventual
proletarianisation of the peasantry.

Marx's analysis of the capitalist penetration of agriculture
has considerable appeal. First, the analogy with industry is a
powerful one, and highly suggestive during the nineteenth century
when new forms of industrial concentration appeared which were
superior to anything that had developed previously. Second, the
proposition that proletarianisation is the ultimate, rather than the
immediate, trajectory of the peasantry, is difficult to refute
empirically, relying as it does on processes which have not fully
matured. For example, Engels was able to argue after Marx's death,
in The Peasant Question in France and Germany, that although the
peasantry was ultimately doomed, its ragged battalions could, in the
interim, be enlisted in support of the working class.

These theoretical strengths should not lead us to overlook the
weaknesses of Marx's position. A general source of difficulty in
Marxist writing on agriculture during the nineteenth century is that
the literature concentrates on the transformation of peasantries,
relating differentiation processes to inherited property relations
and ownership structures. Like Friedmann (1978a) we would question
the relevance of this problem to discussions of simple commodity
production and capitalised family farming in contemporary capitalist
societies. In particular we argue that distinctions need to be made
between capitals as they relate to agriculture, since the effect of
capitalist appropriation on agriculture is not to introduce
wage-labour as a universal form, as is the case with industry.

SIMPLE COMMODITY PRODUCTION IN DEVELOPED COUNTRIES

In view of the confusion that has surrounded many of the categories
in classical Marxist analysis, it is not surprising that the
discussion of the 'family farm' in mature capitalist economies has
provoked a major effort in the recent Marxist literature. There has
been a considerable attempt to theorise household forms of
production, their internal relations and conditions of existence.

This conceptual effort seeks to provide the theoretical basis necessary to specify the structural position in advanced capitalist societies of commercial farms reliant on family labour. Several approaches to this task, at varying levels of abstraction, can be discerned, although all formulations attribute the reproduction and stability of family labour enterprises to their subordinate integration in the circuits of capital. These formulations include attempts to conceptualise simple commodity production as a separate theoretical and empirical category, notably by Harriet Friedmann (1978a,b, 1981), whose work is considered below. Although it owes something to Chayanov, this formulation of simple commodity production is opposed to articulation analyses predicated on the 'peasant' characteristics of rural household activities, particularly the production of use-values for internal consumption. Friedmann's work is distinguishable from that of most other Marxist writers on agriculture in that it is principally concerned with mature capitalist economies and, in the North American case, economies that have no history of 'peasant' production under conditions of land scarcity.

An alternative current found in the literature includes analyses initially elaborated to conceptualise the capitalist incorporation of 'peasant' agriculture in peripheral social formations, and which posit the exploitation of petty commodity producers through mechanisms of unequal exchange and devalorised labour time (Amin 1974, de Janvry and Garramon 1977, Bernstein 1977). In one variant of this structural exploitation approach, the reproduction of family-labour enterprises as a stable social category is characterised as the 'privileged ally' of industrial capital (Vergopoulos 1974, 1978, Buttel 1982). A further extension emphasises the monopoly power of the agro-industrial 'complex' and conceptualises the capitalised commercial family farm as a specific form of the worker/capitalist relation (Faure 1978, Mollard 1978, Wanderley 1979). The static conception of the dynamic of industrial capitals, which derives from the essentialist hypothesis that the reproduction of the peasantry is determined by the requirements of industrial accumulation, is examined in more detail in Goodman, Sorj and Wilkinson (1984).

As in any theoretical undertaking, significant progress and clarity can be achieved by constructing 'ideal types' of theoretical categories before embarking on contingent historical analysis. Harriet Friedmann (1978a,b, 1981) has performed this task with commendable rigour and didactic force in analysing simple commodity production under capitalism. Nevertheless, confidence in the robustness of the theoretical category in question is quickly shaken as we move away from the ideal construct of simple commodity production, predicated on the exclusive use of family labour resources, to historical case-studies of American wheat farming. This confrontation between theory and evidence illustrates several significant limitations of the concept of simple commodity production when applied to family-labour agriculture in advanced capitalist societies.

As her point of departure, Friedmann emphasises that the conditions of existence of simple commodity production presuppose the capitalist mode of production - that is, the full development of commodity relations. 'The logical dependence of simple commodity production, which makes it a form not a mode, is not a general inability to secure its own conditions of existence but a specific requirement of the conditions of existence provided exclusively by the capitalist mode of production' (Friedmann 1981:6). Thus wage labour is required for critical periods of the household's demographic cycle, and it is assumed that all elements of productive and personal consumption are acquired through exchange relations. 'The inclusion of self-subsistence in the definition of simple commodity production is a terminological contradiction' (ibid: 4). However, as we are introduced to the case-study material, it soon becomes apparent that we are dealing with one of those messy 'grey areas' where, despite the very best intentions, theoretical confusion and empirical limitations abound. These concern three issues in particular: wage labour, the nature of the surplus production and class relations.

The 'pure' or ideal case of simple commodity production is predicated on the central, virtually exclusive role of self-employed household members in meeting the technical requirements of labour. These requirements, which are enforced by competition, correspond with the household labour supply, so that 'simple commodity production has <u>no class relations within the enterprise</u>' (Friedmann 1981:13). Thus, in Friedmann's view free wage labour has merely an ancillary function in the labour process of the family enterprise. We would argue, however, that this correspondence between the technical and social bases of household commodity production only holds for certain periods of the household's demographic cycle when teenage and young adult sons are available on the farm. At other times, permanent wage labour therefore becomes a structural requirement for the reproduction of the enterprise, which raises the issue of class relations in production. Although Friedmann (1978a) is reluctant to recognise this implication in her conjunctural analysis, at the theoretical level she acknowledges explicitly this dependence on a capitalist labour market.

The question of wage labour is also crucial to the analysis of the objectives or 'logic' of the household enterprise, and hence the process of reproduction. In the absence of wage labour, the simple commodity form of production has no 'structural requirement for a surplus product'. This follows from the fact that there is no separation between ownership and labour, and therefore profit and wage costs, in the internal organisation of the household enterprise. The behaviour and objectives of the enterprise are then defined by simple reproduction, which will be achieved if gross earnings are sufficient to renew the depreciated elements of productive consumption and provide the household with the socially accepted level of individual consumption. The logic of the 'pure' form of simple commodity production is given by what Bernstein (1977) describes as 'subsistence in the broad sense'.

According to this logic 'there are no economic pressures in simple commodity production for expanded reproduction; that is, simple reproduction (or maintenance of production at the existing level) is adequate for each household' (Friedmann 1978a:37).[1] Expanded reproduction is motivated by demographic and cultural factors, and involves the formation of household enterprises with the same scale of production and level of technology. By contrast, in capitalist production, the separation between ownership and

labour imposes the logic of the appropriation of surplus value and
its realisation in expanded reproduction, reinforcing tendencies
towards the concentration and centralisation of production, even at
constant levels of technology.

> In simple commodity production, the combined return to the
> enterprise and to labour generates no comparable tendency
> towards increased scale of individual enterprises. Demographic
> pressures for expanded reproduction, which must accommodate
> competition, thus create a tendency towards 'fission'..., in
> which income generated in one enterprise is used to establish
> a new one on the same scale.
>
> (Friedmann 1978a:88).

While Friedmann's discussion clarifies the difference in the
logic of expanded reproduction in simple commodity and capitalist
commodity production, the distinction she makes is heavily reliant
on the role of wage labour in the internal organisation of the
respective enterprises. This is a source of major conceptual
weakness since, although the wage labour hired through the
capitalist labour market is acknowledged as a condition of
existence of simple commodity production, indeed its sine qua non,
its use is not 'systematic'. At least not in the same way. Friedmann
argues, that the wage relation represents the crystallisation of the
internal relations of the capitalist enterprise. The confusion
arises from the critical importance which wage labour, and hence the
separation of ownership and labour, can assume in the reproduction
of the 'family farm' for extended periods of the household
demographic cycle, when family labour supply is inadequate. Over the
course of this cycle, household commodity producers thus may
oscillate between the 'pure', ideal type, independent of hired
labour, and small capitalist enterprises. These changing
requirements for hired labour emerge clearly in Bogue's account of
farming on the Illinois and Iowa prairies in the later nineteenth
century:

> In the middle period of the family cycle, when a farmer's sons
> were in their teens or early manhood, he might be tempted to
> enlarge his holdings in order to use the available labour fully
> or to prepare for the time when the sons would want farms of
> their own. During the first and third phases of the cycle he
> often had to hire considerable labour.
>
> (Bogue 1986:185).

Recourse to permanent wage labour to secure the reproduction of the household enterprise also suggests that the categoric statement that simple commodity production imposes 'no structural requirement for surplus product' should be qualified. If competitive pressures dictate the employment of wage labour for extended periods of the generational cycle,[2] it will be necessary for simple commodity producers to earn a surplus product as an absolute condition, though not a relative one, of their reproduction. Of course, this surplus is produced partly by the direct labour of the household. Nevertheless, with the permanent use of wage labour, there is a clear division between ownership and labour, and capitalist class relations are present, if not in their archetypal form, within the enterprise. Wage labour confronts the enterprise as a cost and surplus labour is appropriated as surplus value and realised as profit, the specifically capitalist mode of appropriation. The distinction between family labour and wage labour in highly capitalised commercial farm enterprises and its implications for income and cost categories would seem to require more careful exploration.

This omission may well explain the problematic characterisation of class relations between the farm owner or tenant and his hired labour force:[3]

> Permanent wage labourers, with no kinship connection to the means of production, nevertheless bear a different relation to their employers when the latter are themselves the main source of labour. These no doubt constitute class relations, with some kind of exploitation, but they are different from capitalist relations of production proper.
>
> (Friedmann 1978a:96, emphasis added).

However, the limited differentiation of tasks within the agricultural labour process, represented by the combination of self-employment and free wage labour, should not be allowed to obscure the existence of capitalist class relations. As Marx emphasised, 'only with a minimum capital does the capitalist cease to be a worker himself and (begin) to concern himself entirely with directing work and organising sales' (1976:1027). The question is not whether the employer is also the main source of labour, but whether the employer's essentially entrepreneurial role is altered by the fact that he also contributes labour to the enterprise. As we argue below, the archetypal or 'superior' enterprise in modern agriculture does not necessarily employ a labour process characterised by wage labour relations. However, it is obscurantist to deny class relations when ownership and labour are divorced and, by admission, not mediated by kinship. In this respect, Friedmann's position may be unduly influenced by the discussion of paternalism and personal loyalty, and the possibility that the wage workers in question may be farmers' sons attempting to scale the first rung on the 'agricultural ladder'.

The points referred to above reveal the limitations of the concept of simple commodity production as it has been employed in the analysis of commercial agricultural enterprises based on family labour. These doubts echo those raised in the parallel discussion of the urban 'informal' sector (Bromley and Gerry 1979), which questioned attempts to impose an internal consistency on simple commodity production in order to buttress its use as a conceptual and empirical category (Scott 1977). The danger is one of conferring on simple commodity production the status of a theoretical concept, whereas it is an historically contingent phenomenon, which consequently can be expected to undergo significant transformation and variation in the course of capitalist development. This would explain the difficulties encountered in characterising simple commodity production in empirical analyses. It is further suggested by the subordinate status of simple commodity production under capitalism (and other 'epochs'), so that a continuum of intermediate states may exist between the 'pure' form and archetypal capitalist enterprise (Scott 1979). 'It is important to bear in mind that variations in small-scale production are analysed by Marx as stages in an historical process rather than as separate theoretical states. Modifications in their internal structure occur as a result of the different ways in which they are subjected to capital' (ibid: 111). This subsumption has taken on highly specific characteristics in agriculture, with critical implications for the rural labour process and capitalist accumulation.

THE SUBSUMPTION OF THE RURAL LABOUR PROCESS

The characterisation of commercial farms reliant on family labour as small capitalist enterprises is based not only on the relative importance of wage labour but also on considerations arising from the subsumption of labour by capital. Subsumption in agriculture presents some distinctive features which have not been adequately considered in conceptualising its incorporation by capital. While it is important to analyse the relations of exploitation at the 'point of production', the transformation of the immediate labour process is determined by the movement and competitive struggle of industrial capitals. Recent theoretical formulations have concentrated on the internal structure of the family labour enterprise, neglecting to analyse the determinants of the changing technical basis of production. For example, in Friedmann's analysis, the conjunction between the labour requirements imposed by competition, the prevailing technical conditions of production and the labour supply of farm households, is treated as fortuitous. A more rigorous treatment of the economic and technological context of simple commodity production is required. If accepted such a position would strengthen the view of simple commodity production as an historically contingent form of production. At the same time the theoretical discussion would be strengthened by removing specific historical cases from the forefront of analysis. Friedmann's contribution to our understanding of the simple commodity producer needs to be elaborated in a way that recognises changing economic and technological conditions.

In the following discussion, we consider how to overcome this omission and sketch a tentative approach to the problems presented by agricultural technology and subsumption of the rural labour process. This helps to clarify the reproduction of small capitalist producers on the basis of the inherited labour process. Broadly, the point of departure of our analysis is that agriculture be considered much as any other sector in that capitals seek to valorise all productive activities and fully exploit technological innovations to maximise surplus value. This dynamic leads to the real subsumption of economic activity, which requires the imposition of a capitalist labour process. That is, the achievement of capitalist control and concentration of the means of production, thereby establishing the necessary conditions for expanded accumulation and the rapid advance of the productive forces.

In Marx's analysis of the emergence of the crystallised form of capitalist enterprise, he draws a fundamental distinction between the formal and real subsumption of labour. In the early periods of capitalist development, surplus value may be appropriated as absolute surplus value on the basis of the inherited labour process and without transforming the technical conditions of production. 'Capital subsumes the labour process as it finds it, that is to say, it takes over an existing labour process developed by different and more archaic modes of production' (Marx 1976:1021). With little or no advance in the technical basis of production, it is the 'mode of compulsion' that has changed with the penetration of capital. Marx defined '...the takeover by capital of a mode of labour developed before the emergence of capitalist relations' (ibid:1021) as the formal subsumption of labour under capital. Ceteris paribus, the worker would still be able to undertake the labour process as an independent producer. The transition to a specifically capitalist labour process occurs when capital revolutionises 'the real nature of the labour process as a whole' and extracts surplus value as relative surplus value. This transition, which in industry typically would be associated with a more complex division of labour and large-scale production, involves the real subsumption of labour since the new labour process is beyond the capacity of workers operating as self-employed producers. For Marx, capitalist production can be based equally on formal or real subsumption.

This brief digression is important for the analysis of family labour forms of agricultural production under capitalism since, in most instances,[4] the subsumption of labour is formal. The direct producer in agriculture, by and large, retains control of the labour process and understands the technical basis of production. Yet capitalist development has brought a series of mechanical, chemical and biological innovations. Were farming to remain more or less unchanged by the development of capitalist industrialisation we could expect the 'mode of compulsion' to resemble that set out by Marx in The Eighteenth Brumaire. Family producers would be squeezed

by usurers and the exactions of a state bureaucracy until, through
indebtedness and a failure to meet the economic requirements of the
industrial sector, they would be forced to sell up and join the
wage-labour force. That this has not happened, in Europe or the
United States, is clear from the fact that family labour-based farms
still retain their numerical ascendancy. The capitalist labour
process that Marx observed in large-scale industry was not
transposed to agriculture.

The fact that the transition to a specifically capitalist
labour process in agriculture has not been effected should cause us
to consider to what extent formal subsumption is an obstacle to the
realisation of value in agriculture. The agricultural production
process in Western Europe or North America today is no longer quite
as capital found it.[5] The historical inability of capitalism, and
more particularly agro-industrial capitals, to subsume and
revolutionise the rural production process as a unified whole,
suggests that we should examine the objective constraints on this
process, together with changes in the organisation of industrial
capital during this century as it relates to agriculture.

The principle objective constraints to the imposition of a
unified, specifically capitalist labour process, and hence to the
capacity to revolutionise the means of production, are organic
nature, land and space.[6] As elements of this process become amenable
to industrial reproduction, they have been appropriated by
industrial capitals and reincorporated in agriculture as inputs or
produced means of production. These partial, and historically
discontinuous, appropriations define the origins and development of
agro-industrial capitals and the 'complex' of equipment, processing,
seeds and agri-chemical firms. Compare, for example, the transition
in textiles from the putting-out system to concentrated, mechanised
factory production. Our point here is that in the case of textile
production, a single, unified capitalist labour process was
established in a relatively brief period under the centralised
control of a fairly homogeneous fraction of industrial capital.

The capitalist development of agriculture can be conceptualised
as the competitive movement of industrial capitals to create sectors

of valorisation by re-structuring the inherited, 'pre-industrial'
rural labour process. Industry has progressively appropriated
activities related to production and processing which at earlier
conjunctures were regarded as integral elements of the rural
land-based production process.[7] It is precisely at this stage, in
the industrial sectors created by these appropriations, that
previously 'rural' activities undergo real subsumption to capital,
removing barriers to accumulation and expanded production. One
empirical reflection of these tendencies is the rising share of
industrialised farm inputs, product processing and distribution in
the value of agricultural production.[8] Furthermore such sectoral
value added data do not reveal the full extent of real subsumption
since they exclude production which is no longer rural-based, as in
the case of synthetic raw materials and food substitutes.

In this sense, the survival of farms, with non-reproducible
land as the material base of production, is the measure of the
(current) limits of real subsumption. With this constraint, the
long-run tendency of capitalist development in agriculture is to
undermine the significance of land in production (Murray 1978). We
need only consider the land-saving, productivity-enhancing effects
of hybrid seeds, fertilizer-responsive plants, and the implications
of such biotechnological innovations as recombinant DNA methods and
single cell protein, to appreciate this tendency. Nevertheless, land
and organic nature continue to defy the imposition of a unified,
specifically capitalist production process in agriculture. The
direct rural producer thus retains considerable control of the
labour process, coordinating or managing the integration of those
elements which have 'resisted' real subsumption and the series of
partial appropriations represented by agro-industrial inputs of
equipment, seeds, fertilizers and pesticides (Goodman, Sorj and
Wilkinson, 1984). With this approach to real subsumption, we can
discard unsatisfactory formulations which treat agriculture as a
'special case', whether emphasising the 'peculiarities' of the
labour process, the advantages of 'family'-based production, or rate
of profit arguments.

When we affirm that the long-run tendency of capital is to
transcend the constraints presented by organic nature and land to

its expanded reproduction, we are interpreting the concept of real subsumption in a way that is consistent with the theory of capitalist development. In agriculture, the emergence of the wage relation or the rise of commercial, family labour-based units do not signify real subsumption, freeing capital to transform the technical basis of production. In our view, the real subsumption of agriculture is not to be observed at the 'point of production' of the farm. Rather, it is represented by the long-run tendency of capital to eliminate the labour process as a 'rural' or land-based activity.

Conceptualisations of family labour-based farms couched in terms of their ability to survive under capitalism thus require some reformulation. This follows directly from recognition of the fact that industrial capital, unable to eliminate land and nature as the basis of rural production, has sought valorisation through partial appropriations. However, this is not to contend that agro-industrial capital has met an obstacle in the family farm itself. Such a position would imply that although capital has made major incursions into farming, the family farm has remained unchanged. Rather, 'capital' has been forced by objective limits on its capacity to transform the production process, to pursue a strategy of partial industrial appropriations. Furthermore, changes in the technology of farming, which were expected to confer benefits on large-scale producers in the nineteenth century[9] have conferred no decisive technological advantages on wage-labour enterprises. Indeed, as agro-industrial capitals created further areas of valorisation by partial appropriations of the rural labour process, conditions for surplus realisation by large, wage-labour farms were progressively eroded. In short, technological advance has singularly failed to confirm the classic prognosis of the superiority of capitalist forms of production in agriculture, where these are conceptualised in terms of wage relations. With this perspective, it is not necessary to accept that the family labour-based farm is the permanently privileged partner of agro-industrial capitals, to concede that this form of production is well-adapted to the prevailing patterns of surplus appropriation and technological innovation in agriculture.

IDEOLOGY AND THE LABOUR PROCESS IN WEST EUROPEAN AGRICULTURE

In the preceding section we were at pains to argue that the main
obstacle to the full realisation of value through real subsumption
is not the existence of family-owned farms per se, but the objective
limits which nature and land presented to expanded reproduction in
agriculture. This distinction is important precisely because of the
ideological nature of the discussion of family farming. Most of this
discussion has averred that family farming survives because of its
'family' basis, evoking values such as private property ownership,
thrift and even 'democratization' (Buttell 1983:88). This
ideological account has run parallel with capital's progressive
appropriation of activities once regarded as essential to farming.
Friedmann's work, like our own previous writing on the subject
(Goodman and Redclift 1981), can be criticised for failing to
incorporate these ideological considerations fully into the
analysis.

The capitalised family producer has sought, and in large
measure achieved, political legitimacy by seeking to distance
himself from 'capitalism' at the ideological level, while fully
embracing it at the economic level. A number of factors have
contributed to this very considerable achievement, and together help
explain the direction of agricultural policy in Western Europe.

First, centralising and urbanising influences in most rural
areas have helped create more effective forms of collective action
by producers, although 'class' in a conflictual sense, is a term
which has been studiously avoided. In the French case, and this
remains the single most important case in Continental Europe,
farmers 'obtained new points of comparison...they learned how to
complain' (Tracy 1982:185).

The centralisation and urbanisation of social relations within
rural areas has effectively reduced what Marx regarded as almost a
defining characteristic of the 'peasantry' - its inability to
organise in defence of its own interests. It has been assisted by a
secular decline in the numbers employed in agriculture, and an even
more pronounced decline in the numbers exclusively employed in
agriculture. Changes in tenancy relations, and the virtual

disappearance of sharecropping arrangements also served to give
further ideological saliency to freehold land ownership.

Even before most of these demographic and structural changes
were effected, large farmers' organisations disguised their class
interests behind the myth of a single class: the classe paysanne,
in the case of the Societe des Agriculteurs de France. In England,
the threat represented by the organisation of agricultural workers
helped in establishing the National Farmers' Union and the Country
Landowners' Association at the beginning of the century. Both these
organisations were to become skilled at identifying the sectional
interests of large farmers with farmers in general and, at another
remove, with the nation.

The association of national patriotism with the agricultural
interest is another factor of importance in the legitimation of
European farmers' groups. At one level this association can be
traced to the immediate effects of two World Wars, in which national
security became identified with agricultural self-sufficiency, as in
the British case. In France, the aftermath of the First World War
created conditions favourable to greater state intervention in the
domestic market and helped sustain tariff protection. Throughout
Europe the rising tide of agrarian populism gathered many new
adherents to its nationalist, anti-communist banner during the
1930s. It also ensured that the governments of the time could not
ignore 'Europe's Farm Problem' when peace returned a decade later.

Since the last War the changing ethnic composition of many
European cities has also been a potent factor in the way rural areas
are perceived. By 1981 there were twelve and a half million
'guest-workers' in the E.E.C. countries, as well as millions of
black immigrants from ex-colonial countries. Against the backdrop of
rapidly changing cities, the European countryside, from which many
farming families had migrated within the previous two or three
generations, came to be seen as a repository of cultural continuity.
Populist ideology emphasised both the non-capitalist nature of
family farming and the deep historical roots of the European
'peasantry' to good political effect. Few questioned the validity of
either proposition.

A third ideological factor which has helped to legitimate the
West European farming lobby is best described as the opposition to
economic growth, as associated with urban industrial development.
During most of the post-War period European farmers have sought to
associate themselves with the preservation of nature in the face of
increasing urban pressure. Not until the last decade have the
negative environmental effects of modern agriculture been accorded
any weight. In Britain the association of farming with the
conservation of the countryside was reflected in planning
legislation, which discouraged 'urban' development in rural areas
while enabling farmers to benefit from increasing growth in
agricultural production. Social responsibility was to be exercised
by negative sanctions against 'urban' developers rather than by
planning controls on farmers like those which the rest of the
population recognised. The situation in France was similar, although
anti-productivist sentiments were much more common (Bergmann 1983),
and the rural vote could be secured by a simple appeal to the
economic self-interest of the farmer.

If we consider the ideological level as part of the labour
process, in the development of capitalised farm enterprises, it
becomes easier to appreciate the Common Agricultural Policy's hold
on such a large part of the budgetary resources of the European
Economic Community. The political conditions under which different
capitals have been able to appropriate a bigger share of West
European agriculture were established in the Treaty of Rome, but the
ideological resonance of the 'family farm' has a longer, and even
more pervasive, history.

FOOTNOTES

1 Friedmann here is adopting the method of Banaji (1977) and distinguishing relations of exploitation, such as wage labour, at the 'point of production' from relations of production and the laws of motion, which characterise different modes ('epochs') of production.

2 This pressure on family-labour farmers is noted by Bogue (1968) in the later nineteenth century when he observes that 'It is also clear that the farm operated by one owner-operator...was an inefficient operation in a mechanical age, where many operations could be most effectively performed by two men' (184-5, n.34).

3 When wage workers have kinship relations to an owner of the means of production, we can accept Friedmann's point on their class position. Yet analytically this does not change the role of these workers in the labour process nor the antagonistic interests between them and their current employers.

4 Real subsumption is most closely approximated by certain forms of outgrower relations, particularly in poultry and livestock production utilising hybridisation techniques, since the farmer-worker retains virtually no control over the technical conditions of production.

5 The marketing, processing and packaging of food contribute more to GNP and employment in developed economies than does the agricultural sector itself. In addition the price paid by consumers for food is on average three times the price received by farmers (International Broadcasting Trust 1983:p.7).

6 The significance of 'space' in rural production is exemplified by the settlement of whole continents in the nineteenth century on the basis of the rising European demand for staple grains and meat. Conversely, while cotton textile production for an expanding world market was soon concentrated in South Lancashire, the cultivation of its raw material occupied vast areas of the globe.

7 This approach is diametrically opposed to the notion of a static division between 'agriculture' and 'industry'. 'Agro-industry embraces a constantly changing mix of capitals and represents the continuous attempt by industrial capitals to transform agriculture into an industrial process. As such it has no static, pre-established limits'. 'In this respect, the agro-industrial 'complex' represents an incomplete, transitional phase in the industrial appropriation of agriculture'. (See Goodman, Sorj and Wilkinson 1984:196 and also Redclift 1984:118-121.)

8 In France purchased inputs absorbed 37 per cent of the value of production in 1980. In 1981 the total debt of French farmers was nearly 150 per cent of value added (close to that of West Germany and the Netherlands). Bergmann 1983:271.

9 For example, the discussion between David, Kautsky and others
  within the German Social Democratic Party in the 1890s. The
  family labour farm's survival at the time was largely attributed
  to the family's ability to produce its own means of subsistence,
  a capacity which specialisation removed from larger producers.
  (see Hussain and Tribe 1984.)

REFERENCES

Amin, S. 1974. Accumulation on a world scale. New York: Monthly
    Review Press.
Banaji, J. 1977. Modes of production in a materialist conception of
    history. Capital and Class 3, 1-44.
Bergmann, D. 1983. French agriculture: trends, outlook and policies.
    Food Policy 8, 4, 270-286.
Bernstein, H. 1977. Notes on capital and peasantry. Rev. African
    Pol. Econ. 10, 60-73.
Bogue, A. G. 1968. From prairie to cornbelt. Chicago: Quadrangle
    Books.
Bromley, R. and Gerry, C. 1979. The casual poor in third world
    cities. Chichester: John Wiley.
Buttel, F. 1982. The political economy of agriculture in advanced
    industrial societies. Current perspectives in social theory 3,
    27-55.
Buttel, F. 1983. Beyond the family farm. In Technology and rural
    social change. 87-107. (G. F. Summers (ed.)) Boulder: Westview
    Press.
Faure, C. 1978. Agriculture et capitalisme. Paris: Anthropos.
Friedmann, H. 1978a. Simple commodity production and wage labour in
    the American plains. J. Peas. Stud. 6, 71-100.
Friedmann, H. 1978b. World market, state and family farm: social
    bases of household production in the era of wage labour.
    Comp. Stud. Soc. Hist. 20, 545-586.
Friedmann, H. 1981. The family farm in advanced capitalism: Outline
    of a theory of simple commodity production in agriculture.
    Toronto: mimeo.
Goodman, D. and Redclift, M. 1981. From peasant to proletarian:
    capitalist development and agrarian transitions. Oxford: Basil
    Blackwell.
Goodman, D., Sorj, B. and Wilkinson, J. 1984. Agro-industry, state
    policy and rural social structures: recent analyses of
    proletarianisation in Brazilian agriculture. In
    Proletarianisation in the third world. 189-215. (B. Munslow and
    H. Finch (eds.)) London: Croom Helm.
Hussain, A. and Tribe, K. 1984 (eds.). Paths of development in
    capitalist agriculture. London: MacMillan.
International Broadcasting Trust, 1983. Utopia Limited. London: IBT
    Education.
de Janvry, A. and Garramon, C. 1977. The dynamics of rural poverty
    in Latin America. J. Peas. Stud. 4, 206-216.
Marx, K. 1968. The Eighteenth Brumaire of Louis Napoleon. In
    Selected works. (K. Marx and F. Engels) London: Lawrence and
    Wishart.
Marx, K. 1972. Capital, Vol. III London: Lawrence and Wishart.
Marx, K. 1976. Capital, Vol. I Harmondsworth: Penguin.

Mollard, A. 1978. Paysans exploites. Grenoble: Presses Universitares de Grenoble.

Murray, R. 1978. Value and theory of rent: part two. Capital and Class 4.

Newby, H. 1977. The deferential worker. London: Allen Lane.

Pearse, A. 1980. Seeds of plenty, seeds of want: social and economic implications of the green revolution. Oxford: Clarendon Press.

Redclift, M. 1984. Development and the environmental crisis: red or green alternatives? London: Methuen.

Scott, A. 1977. Notes on the theoretical status of petty commodity production. Paper presented to the Development Study Group of the British Sociological Association, mimeo.

Tracy, M. 1982. Agriculture in Western Europe - challenge and response 1880-1920. London: Granada.

Vergopoulos, K. 1974. La question paysanne et le capitalisme. Paris: Anthropos.

Wanderley, M. de N. B. 1979. U campones: um trabalhador para o capital, (Texto para discussor, no.2. Capinas, UNICAMP/IFCH/ DEPE), mimeo.

3 Family enterprises in agriculture: structural limits and political possibilities

HARRIET FRIEDMANN

INTRODUCTION

The family farm in advanced capitalist economies is a 'problem' that just will not go away. All of classical sociology, including Marxism, expected the household to give way completely to 'modern' industry. Those for whom it is not a problem have narrowed the classical vision: including those economists who are unconcerned with relations of production within enterprises, and those rural sociologists who are uncritical of the ideological 'naturalness' of family farming.

In considering the role of family farming within capitalist agriculture we must first address the more general question of what is specific to family enterprises in capitalist economies. We shall then be in a position to enquire into the place of simple commodity production within the commodity chains linking farms eventually to consumers of food, and into the politics surrounding food and agriculture. However, before embarking on this course, it is necessary to lay to rest another line of enquiry which misconceivedly would distinguish agriculture from other types of production in capitalist economies.

THE LACK OF SPECIFICITY OF AGRICULTURE IN CAPITALIST ECONOMIES

Various characteristics have been put forward as marking off agriculture but on closer scrutiny none is unique to it. For example, agriculture can be understood as a 'rural production

process' which uses land as a 'material base of production' as
claimed by Goodman and Redclift in this volume. Yet any conceivable
production appropriates nature and takes place in space. Every
product of labour ultimately can be traced back to natural elements.
Capitalism changes the significance of space by systematically
attaching prices to specific bits of the earth's surface, and even
to what is above and below the surface; and the price per unit of
area - rent - is no less important to urban property developers than
to farmers.

Agriculture is also the production of plants and animals as use
values. These have historically been produced using land extensively
relative to other production processes. No one has yet redefined the
word to deal with production of animals 'on concrete' (Newby 1980)
or of plants in glass houses or even in laboratories. Nor can the
term accommodate the tourist 'industry', which requires that the
land, and sometimes the plants, animals, and even the people on it,
appear picturesque. Thus land gives no unity to agriculture.

Agriculture takes various forms: the marginal (to capital)
production of plants and animals, at least partly for direct
consumption or local markets ('peasants'); part-time farming by
people who are waged workers, capitalists, professionals, managers,
etc. in the larger economy; simple commodity production, which is
fully integrated into specialized product markets (the 'family
farm'); capitalist production employing temporary, marginal, or
seasonal labour; and capitalist production undertaken by
multinational capitals and employing stable, often unionised, labour
forces ('industrial agriculture'). These differing forms are often
seen as part of a continuum and as a tendential chronology (Goodman
and Redclift, chapter 2).[1] As I argue below, I think that this is
wrong. Agriculture does not derive unity from labour.

Although I shall continue to use the word to refer to the
production of plants and animals on the land, I want to emphasize
that it is folly to try to characterize 'agriculture' as such.
Despite our commonsense ideas and impressions, capitals producing
specific use-values structure and restructure the forms of
production, including combinations of land and labour, in relation

to complex and changing determinants of profitability. For plants
and animals as for other products, this sometimes means direct
organization, including formal or real subordination of labour (and
land). Sometimes, however, capital indirectly relates to other forms
of production, such as family farms, for which opportunities open
and close.

It is convenient but not entirely accurate to talk historically
of capitalism as separating industry from agriculture.
Proto-industry, often combined with agriculture, was a widespread
and longlived aspect of the formative period of capitalism. The idea
that capitalism separated industry from agriculture describes the
eventual, but probably not ultimate, specialization of industrial
and agricultural activities in town and country. Suburban and even
rural industrialization is the recent counterpart to the virtual
disappearance of traditionally working class occupations and
residences from urban centres (Zukin 1982). Separation, therefore,
is not an adequate starting point for the analysis of agriculture
under capitalism, especially in the late twentieth century.

It is crucial that we not confuse 'industrialization' with the
development of capitalism. It is tempting and easy to do so,
especially when the issue is the 'real' subordination of labour. A
counter example comes from Mintz (1985:51-52), who persuasively
describes the sugar-producing slave plantations of the West Indies
as 'industrial enterprises', but not capitalist:

> What made the early plantation system agro-industrial was the
> combination of agriculture and processing under one authority:
> discipline was probably its first essential feature. This was
> because neither mill nor field could be separately
> (independently) productive. Second was the organization of the
> labor force itself, part skilled, part unskilled, and organized
> in terms of the plantation's overall productive goals. To the
> extent possible, the labor was composed of interchangeable
> units - much of the labor was homogeneous, in the eyes of the
> producers - characteristic of a lengthy period much later in
> the history of capitalism. Third, the system was
> time-conscious. This time-consciousness was dictated by the
> nature of the sugar cane and its processing requirements, but
> it permeated all phases of plantation life and accorded well
> with the emphasis on time that was later to become a central
> feature of capitalist industry. The combination of field and
> factory, of skilled workers with unskilled, and the strictness
> of scheduling together gave an industrial cast to plantation

enterprises, even though the use of coercion to exact labor might have seemed somewhat unfamiliar to latter-day capitalists.

A modern example, which is both industrial and capitalist, is California lettuce production. Thomas (1985, 1981, also Friedland, Barton, and Thomas 1981) uses the term 'industrial agriculture' to denote the technical division of labour creating the 'collective worker'. Like Adam Smith's original pinmaking example, it is independent of any technological change. Teams of three men pick and pack, and productivity inheres in the team. The labour force is stable and unionized. As to land, the multinational firms which organize production rent land throughout the West and Southwest so that they have continuous production by moving their permanent labour forces to follow the climate. Thomas uses 'industrial agriculture' to make a point about the continuity of labour processes across 'sectors'.[2] What he is describing is fully capitalist organization, significantly subordinating both labour and land to routine production, but without machinery of any importance.

The answer to the question, what is specific to agriculture under capitalism, is nothing. It is not that agriculture has developed parallel to industry, but that links in complex chains connect production of specific plants and animals to equally specific manufacturing processes. Instead of contrasting agriculture to industry, and thus assuming the conceptual unity of each, I think it much more useful to begin with the idea that productive capital organizes the production of specific plants and animals, often on the land and often bound by the specific requirements of soil and climate. These production processes involve specific social relations within, across, and among enterprises.

WHAT IS SPECIFIC TO FAMILY ENTERPRISES?

In an economy based on wage relations, the organization of labour through a family/household complex is clearly distinct from the typical capitalist enterprise. Similarly, the unity of property and labour is distinctive. Capitalism, by definition, separates property and labour, with the monopoly of the means of production on one side and the sale of labour power on the other. Yet family enterprises continually arise and persist, as well as decline, in capitalist economies.

Two things, therefore, are specific to family enterprises in capitalist economies: the labour process and property relations. First, production is organized through kinship and there is a division of labour by gender and age. This social organization is remarkable in an economy of generalized circulation, where labour markets are a central fact of life for everyone. Despite alternatives both to family labour for the enterprise, and to working in the family for its members, these units emerge periodically - and often persist tenaciously - in various spheres of production and commerce.

Second, property and labour are combined. This too is remarkable in an economy premised on their general separation and recombination through the purchase of labour power by capital. Ownership is not equal, of course, any more than participation in the direction of the enterprise or in work is equal among family members. But the basis of inequality is to be found in gender and age relations. They are subject as much to laws governing family (personal) property and the rights of men, women, and children, as to laws governing productive (often corporate) property and relations between employees and workers.

Family production, in the broadest sense, has an ancient lineage. Even in the narrower sense of a family/household complex it is very old and various. Yet it takes on distinctive features because of both the dominance of commodities and specific demographic characteristics of capitalist societies. The precapitalist 'master' of the productive household (Laslett 1965) gives way in the capitalist economy to a combined 'head of household' and 'director of enterprise'. There is no necessary continuity implied in the comparison, for family enterprises exist in all capitalist economies, including those with little precapitalist history, and among many cultural groups within them, including immigrants adapting their practices to new contexts.

The dynamics of family enterprises stem from the interplay of household and business. This is perhaps most clear in cases of

succession to a new generation. There are two processes by which a
new generation can be accommodated: 'fission' or creation of a
second enterprise; and 'expansion' to incorporate the labour of two
generations. The path taken by any particular family enterprise
depends on both external and internal factors. In addition to
markets for the product of the enterprise and the availability of
capital, the specific relations within the family are crucial. On
one side, the desire by sons (generally) to succeed too soon, or for
more than one child to inherit, may clash with parental wishes and
power, and with enterprise constraints. On the other, parental
desire to perpetuate the family enterprise may clash with desires
for independence or immediate incomes on the part of the next
generation (Lem forthcoming).

Even within the working life of one generation, the family
enterprise is subject to double pressures from capitalist society -
on the family and on the enterprise. The strain shows. Capitalism
marginalizes the family (Briskin 1985).[3] Property based on the
family is therefore fragile. As marriage becomes increasingly
impermanent, those families able to assume permanence and act on it,
often immigrants from less developed economies, are better placed to
establish and sustain family enterprises. Of course, the reverse
causation is also at play. Families with property but little to
translate into marketable skills as individuals on the labour
market, have unusual (or unusually equal) incentives to remain
together.

What is _not_ specific to family enterprises is wage labour. It
is not surprising that they resort to the labour market for
additional labour. That is a general phenomenon - even private
households pay people for work, for that is how labour and
consumption are mediated in capitalist economies. When labour is
employed productively, however, it is different. Of course, it
allows family enterprises to function despite the variation in
available labour during the family life cycle (Friedmann 1978b,
Bogue 1968). To this should be added the variation in interests and
aspirations among family members, which vary across enterprises,
regions, cultures, genders, and generations. When wives do not wish
to participate in the enterprise, or children do not wish to take

over, and when cultural and legal sanctions permit, the choice is to employ labour or to fail. However as long as property and labour are combined, through the owner continuing to work in production, a transition has not yet occurred.

To be sure dependence on hired labour is a problem for the reproduction of family enterprises. Such dependence is as dangerous as it is necessary, since it may as easily undermine the enterprise entirely as enhance its transition to a capitalist form of production. For example, purchase of additional equipment may involve an irreversible commitment to wage labour, which could prove disastrous if market or other conditions change. What is specific about wage labour in simple commodity production is the peculiar and contradictory role that it plays: enterprises must draw on the labour market of the larger capitalist economy, yet are vulnerable on several fronts to its pressures.[4]

FAMILY FARMS IN CAPITALIST ECONOMIES

The larger and less divisible the property as a production unit, the more complex is this interplay. It has been most meticulously described for family farms, in which the acquisition of land, equipment, and sometimes livestock represents a major, long-term commitment to production for volatile markets. A sustained research project headed by anthropologist John W. Bennett (1982) studied a farming community on the Canadian plains for approximately twelve years during the 1960s and 1970s. With biographical interviews they were able to reconstruct the history of farms and farm families - what they call 'agrifamilies' - over three generations, which usually included the original homestead. One of Bennett's collaborators, Seena B. Kohl (1976), was particularly concerned to distinguish between the perceived and actual roles of women in family farms in the community they call Jasper.

Family farms are both competitive enterprises, frequently in highly volatile world markets, and sources of collective income for families. In this sense, the enterprise may be thought to have 'needs' for investment, just as the family has consumption needs. In contrast to capitalist enterprises, the same group invests and consumes, since there is no structural division between profit and wages. There is therefore a trade-off between investment and consumption (Friedmann 1978a:559-63), or in systems language, 'between production goals and household maintenance goals' (Bennett 1982:134). Under the rigours of the market, the family's needs are a charge against the enterprise (ibid:115), different from a wage bill in that they are flexible, even to the point, in Chayanovian terms, of 'self-exploitation'. Women are responsible for managing the household under this constraint, so that domestic labour becomes crucial to the enterprise. Their ideology supports this practice, as revealed in the following statement from a Jasper woman (quoted in ibid:166):

> When it comes to spending a dollar I always ask myself, What is it going to make?... If you buy a rug, well you have a rug, but if you use the same money and buy a cow, the cow has calves and with the calves you can get the rug but you still have the cow!

As in the case of precapitalist households, the unity of kin group and work group creates dual (at least) roles for each member: father/boss - manager; mother - domestic worker - childcare/accountant - personnel manager - sometime labourer - purchasing agent - public relations; child/labourer - domestic worker. Because of this mix of structures and roles, formal rationality is difficult to sustain. Regardless of the decision-making process within households, and of the tendency in varying degrees to favour the enterprise over family consumption, the seamless division of productive and personal consumption and the fluidity of family labour potentially and actually available, prohibits full calculation on behalf of the enterprise (ibid:118, also Mooney 1985).

'Agrifamilies', as Bennett and Kohl (ibid:140-47) call them, have intertwined cycles of household and enterprise. The authors describe the stages of family life and enterprise development for a typical successful farm in southwestern Saskatchewan. Between 1910 and 1930 there was a 'period of enterprise', in which the farm was established and the family begun. In the first decade the bachelor took out a homestead and made initial improvements, at the same time working as a labourer and saving his wages. He then married, bought

more land and had a first child. During the second decade, two more
children were born and the first became old enough to work on the
farm; the enterprise became viable through additional investment in
land, buildings and machinery. The period 1930-40 was a time of
crisis, in which survival depended on government assistance and
wages from off-farm labour, while all the children worked simply to
maintain the farm. 1940-60 was the 'intensive development phase', in
which there was increased scale of enterprise and intergenerational
succession. More land was purchased, and the enterprise was shifted
onto a more profitable basis as a specialized cattle operation
instead of grain farm. This included starting a herd, joining a
cooperative and then a community pasture, and irrigating. At the
beginning of this period the first son married and younger siblings
left the farm. The two households lived together and worked the farm
for several years until the retirement of the parents, who were then
paid a monthly income by the son, now head of the new farm family.

These combined cycles pose problems specific to the farm family
and to the family enterprise. For the family, there are problems
associated with supervision of the children's labour by the father,
of deferred consumption, of occupational choice affecting whether
succession occurs, by whom, and when, of shared home and enterprise
by two families during the transition. For the enterprise, the most
conspicuous feature is its dependence, at all times - but especially
across generations - on family formation and development for its
survival. In Jasper as elsewhere, marriage and establishment of the
enterprise go together, and despite deferred personal consumption,
often for long periods, children are not delayed (ibid:146). They
provide labour within about ten years, about the time that
investments can begin to make use of it. Then they are ready to take
over, perhaps too early for the convenience of the parents, but
generally not too late (see the case studies in ibid:173-81).

At least as important is the understated role of the wife in
family enterprises. The role of the farm wife was more appreciated
at the beginning of the century. Her recent eclipse derives from the
new and socially dominant reality of households devoted exclusively
to consumption underwritten by wages earned outside the home,
preferably by men. Despite the espousal of conventional ideas about

women's dependence in Jasper there is a deeper reality of women's role in the enterprise, at least equal and perhaps even more important than that of men. First of all, household management is more important to the family enterprise than to propertyless families, since what is saved there is available for investment. Second, to the extent that women do assume the ideologically specified 'expressive' role of conflict negotiation and mediation, this is crucial for labour relations, especially between fathers and sons. Third, women do work on the farm: in addition to purchasing, bookkeeping, recording breeding data, financial management, political and community liaison, all done by wives who typically have greater education and social skills than their husbands, increased mechanization in the 1970s has renewed the direct participation of women in technical labour and management. In Saskatchewan as a whole, in addition to the presence of women operators, farm wives spend at least 'half-time' and often more than 40 hours per week on direct farm work. Despite the expressed belief by both sexes that women are 'helpers', they have a double load comparable to that of women who work for wages and at domestic labour, yet without any monetary or symbolic recognition (ibid: 160-71). That women are important in family farms in England, too, is suggested by their reduced involvement during a transition to capitalist organization, in which neither member of the couple works at manual tasks (Symes and Marsden 1983:245); and in the greater tendency of women from larger, than from smaller, farms to hold outside jobs (Gasson 1984:224).

Of course, the division of labour by gender and age also includes individual contributions, both on and off the farm. Bouquet (1982:228) has described the 'transformation of women's work roles as (an)...accompaniment to the emergence of the family farm' in a tourist area in southwest England. This involved a 'commercialisation of the domestic domain', in this case through putting up visitors in the farmhouse. This extends the more frequent small commerce of women, such as selling eggs or garden produce in local markets, into fully developed enterprises, distinct yet inseparable from the farm. According to Bouquet, this strategy allowed these family farms to emerge and prosper, since they could simultaneously finance investment necessary for postwar dairying and increase the level of family consumption. More typically, though, the specialisation of enterprises and the changed markets for farm produce have squeezed out much of the small commerce of family farms.

Family enterprises, then, involve a specific complex of kinship and property. This complex rests on institutions principally directed towards the major forms of property and the family in capitalist societies. For example, in Canada the family law giving the wife rights to the matrimonial home has effects on the family farm: it gives her no rights to the enterprise, and at the same time limits the rights over the enterprise of the (male) owner and, interestingly, his creditors. A more complicated example, which might also apply to nonfarm family enterprises, is the intersection of tax rules, labour codes, and credit practices. In Sasketchewan, the farm wife is not entitled to half the enterprise and is thus not legally an owner. Without title, she cannot receive credit; yet with title, she loses her matrimonial 'homestead rights'. Conversely, it is difficult for her to claim the rights of labourers. Her husband can deduct wages from his income taxes for hired workers and for his own children, but not for her. Without employed status, she is not automatically entitled to workman's compensation in case of injury, or to training allowances from manpower agencies, and so on (Bennett 1982:168).

The problem of labour for family farms is bigger than usual for family enterprises when land use is extensive. Co-operation is one response and this is discussed below. Distance from labour markets can either create labour shortage or restrict alternative employment. In the latter case, the farm family may constitute a captive labour pool. On other farms, the 'hired man' is in an equivalent position. This is a role which may be more akin to the 'servant' or apprentice of the precapitalist household than to modern waged labour. Working and living with a farm family, with partial payment in room and board, are dissimilar from the experience of most capitalist labour and the typical wage relationship. This illustrates the special role employed labour can play in family enterprises.

Another example, with very different social and political
implications, and in response to labour shortage in situations of
labour mobility (rather than immobility), is resort to special
labour forces, often ethnically distinct and subordinate. Even more
than for family labour or the worker-boarder, such labour may be
grueling and the wages low. Benefiting from the political power of
huge agricultural corporations in California, smaller farmers, such
as tomato growers on contract to processing companies, have
benefited in some periods from special state-to-state arrangements
to bring in seasonal foreign workers. The Canadian government does
the same for relatively small farms in Ontario fruit and vegetable
production. Even when they are citizens, migrant workers, such as
Mexican Americans in the United States, travel the country following
the harvest, subsidizing family as well as capitalist enterprises
with their cheap labour (Friedland and Nelkin 1971). Simple
commodity producers in Europe also rely on immigrant workers and on
such subordinate ethnic groups as gypsies for casual or seasonal
labour.

The most interesting alternative to labour shortage is
co-operation. The history of agricultural co-operation in most
countries contains an anticapitalist ideological component, yet the
most successful co-operatives in terms of scale and persistence have
not surprisingly been those most attuned to the capitalist economy.
From marketing boards accorded monopoly powers by states and acting
on behalf of specialized producers, as in England and Canada, to
giant food processing enterprises nominally owned by farmers in
France, the Netherlands and the United States, little is left that
is distinctly co-operative. The large French co-operatives, which
account for a fifth of the turnover of agro-industries in that
country, may have advantages of supply over private capitals, but
they are relatively inflexible because of juridical restrictions on
their management. They have become, in the words of Bombal and
Chalmin (1980:78-80), 'les grandes nébuleuses', which are
nonetheless often difficult to distinguish from capitalist
enterprises, by their employees, by their customers, and to some
extent even by their members.

Yet co-operation can also solve labour problems of family

enterprises, especially where agriculture is extensive. On the
Canadian and American prairies, farmers frequently adopt informal
and sometimes formal exchanges of labour, in addition to pooling
resources for credit and marketing.[5] When everyone gathers round for
a barnraising or, as in a very recent example, when farmers who have
been spared the worst of a drought and infestation send feed corn to
those who are suffering the most, it is clearly informal and
communitarian. Yet as Bennett (1982:116) puts it, 'strong
individualistic and nuclear-family values sometimes inhibit this
growth in adaptive co-operation'. He goes on to compare family farms
in the district he studied with those of religious communities:

> In Jasper, the contrast between the communal - co-operative
> Hutterian colonies and the neighborhoods of small farmers was
> striking: our studies showed that even adjoining farms operated
> by blood brothers manifested less co-operative interaction than
> the nuclear families composing the colony, although the
> non-Hutterian groups were suffering from a shortage of labor
> and other forms of assistance, which co-operation could have
> gone far to alleviate.

Thus co-operation exposes the structural and political
uniqueness of family farms. To analyse this and similar phenomena,
we must draw on a wider conceptual framework. One dimension derives
from demography and family sociology. The other is the theory of
value to interpret the social division of labour. A key concept here
is that of simple commodity production, which refers to the
contradictory unity of property and labour in an economy
characterized by the general circulation of commodities (including,
of course, labour power and rights to land). Of course in reality it
is no more to be found in 'pure' form than is a 'pure proletarian'
or a 'pure capitalist'. As I think I have demonstrated elsewhere,
simple commodity production is historically contingent in a double
sense; like capitalist enterprises, it is specific to the modern
epoch; but unlike capitalist enterprises, it need not exist even in
the capitalist epoch (Friedmann 1978b, 1981).

The unity of property and labour in simple commodity production
is contradictory in capitalist societies because it internalizes
within one person or family the structured conflict between property
owners and labourers, who are usually related as employers and
employees. It is important because it creates contradictory

possibilities for politics and thus for political alliances. As property owners, the principal relation of simple commodity producers is to the market, which they enter as sellers of the products of labour. As direct labourers, they are working to live. For this they do not enter the market, as proletarians do, but work directly with their own means of production. Insofar as this work is central to their consciousness, it is part of the totality of the social relations within which they live, just as consumption of the wage is for proletarians.

Co-operative marketing taps the property dimension of family farming. It does so whether farmers undertake co-operation on their own initiative or rely on a state board. These arrangements are generally part of price support politics, which emphasize and reinforce that aspect of farmers in which they enter markets as sellers of commodities. When farmers co-operate in exchanging or pooling labour, the other aspect is prominent in their practice and potentially in their consciousness and politics. Yet given the dominance of capital in the larger social institutions and ideologies, labour co-operation is not an equal counterpart to market co-operation, and is much more difficult to sustain.

THE FAMILY FARM IN CAPITALIST POLITICS

The ideological interaction between family farming and capitalist institutions is not at all one way. As Bennett (1982:112) puts it, 'the 'family farm' is as much ideological imagery as it is socioeconomic fact, it is a symbol of all that is traditionally sound and moral (or unsound and archaic) in North American rural society'. Yet this comment is the prelude to an extended study of such farms in all their specificity as enterprises based on familial organization, resources, and needs. Far from exhausting the topic of family farming, to point to its ideological resonances is to urge the importance of 'sober factual analysis' (ibid). The success of farmers in North America and Western Europe in using non-capitalist appeals to urge policies in their corporate interest, rests on a material basis. All groups mobilize what ideological arguments they can, few of which stand up to critical sociological scrutiny. Yet all successful ideological appeals refer to some reality, however obliquely. In the case of family farming, the material reality of family at the centre of the enterprise (however one values family) cannot be dismissed.[6]

One direct connection between farm politics and larger social issues, then, concerns women's equality and the family. The connection is rarely made, of course, but this is a challenge. When the issue is joined, as it was several years ago in Canada, feminist organizations seem the only ones concerned, and then in a way that does not distinguish productive from personal property. The Murdoch case in Alberta involved a divorce in which the court awarded the family farm totally to the husband, despite serious violence by him and despite the wife's acknowledged labour contribution in establishing and operating the enterprise for many years. Changes, proposed or actual, within the law of matrimonial property have special, though seldom addressed, implications for family farming and could fruitfully be correlated with other areas of law, such as that governing farm credit and farm tenure (Bennett 1982: 149-50, 202-3).

A second area in which farm politics potentially intersect with larger social issues is agricultural labour. It is usually exempted from the rights granted generally to workers, either explicitly, as in U.S. legislation, or through the legal procedures governing immigrant workers. In the U.S., Mexicans under the now defunct bracero program were brought in for seasonal contracts and then sent home, all through arrangements on behalf of employers between the U.S. and Mexican governments. Similar arrangements now apply to Caribbean workers in Ontario farms. Generally, these workers are in enterprises that are clearly capitalist. Yet various kinds of migrant labour and of immigrant or ethnically subordinate labour support simple commodity production, in Europe as in North America. The conflicts and alliances possible in this situation should be addressed.

Third, farm politics must address ecology and diet. Both impinge on agriculture directly. Farmers do environmental damage under market pressures, sometimes unwillingly. Political restrictions may restrain particular practices, but in a context of international competition they do not necessarily remove market pressures. The issue is analogous to pollution controls on industry, but also includes the shape of the landscape and other aspects of land-use and regional planning (Newby 1980). General pollution controls and planning regulations, however, impinge on farmers as members of rural communities as well as owners of competitive enterprises. There might be a basis for alliance against competitive forces through imaginative politics.

Food politics, especially those emphasizing the consumption of local products which require safer agricultural practices, can create new markets for the products of small farms. These markets have already provided scope for successful small enterprises in food processing ('health foods'), enough so that multinational corporations have begun to buy them up, often without changing their production processes. It is conceivable that these markets could provide the possibility for new farming techniques, which would further encourage reduced scale in agriculture and a sounder relation to the environment (Jackson and Bender 1984). This conjunction of interests might finally supersede the traditional hostility, reinforced by politics focused on prices, between farmers and consumers.

Finally, these specific intersections of farm and other politics develop within political frameworks relating to property. Price support policies in Europe and North America systematically underwrite property: they resemble corporate subsidies more than social welfare supplements to individuals or families.[7] This issue is not unique to farmers or even to what we usually think of as simple commodity production. The professions, for instance, sometimes claim property rights in the same way. Thus doctors (most in the United States and Canada, private ones in England) who insist on payment for each item of service instead of employment even at a very high salary, are claiming property rights - provision of commodifed services under no supervision except by their corporate body, and for prices which, though mediated, are ultimately set by the market.

Such property has long been the ideological basis for general claims of legitimacy for capitalism. Yet the claim that control over conditions of work and income should reside in individuals or local work groups is as valid as the parallel claims for autonomy of simple commodity producers. State control, centralization, and large scale production are now questioned by left as well as right. Farmers are workers and property owners. Co-operation has so far produced giant enterprises resembling capitalist firms more than worker cooperatives. The challenge is to find a way to create democratic, responsible agriculture in harmony with the environment that farmers and their near and distant neighbours can support.

FOOTNOTES

1   Part-time farming fits into the general, and generally implicit,
    developmental scheme after peasants when the subject is
    peasant-workers; when gentleman farming is at issue, it of course
    comes later. Since I reject any such chronology, it does not
    matter whether this is treated as one or two categories.

2   For a critique of the now classic notion of primary, industrial,
    and service sectors, see Walker (1985).

3   In her brilliant analysis of domestic labour, Briskin shows how
    the family is at once marginal to capital and central to the life
    experience of people in capitalist societies. Put differently,
    one cannot deduce the family from the laws of motion of capital,
    though one can identify a space (the sphere of 'not commodity
    production') in which labour power, which cannot be reproduced by
    capital, reproduces itself. This sphere, which is the one in
    which people live, is dependent on wages and on purchasing
    commodities, but it need not take any particular form, including
    kinship generally or the nuclear family in particular. The fact
    that families have always occupied this sphere, and continue to
    do so despite everything, is an historical legacy of great
    importance. The usefulness of this formulation is that it tells
    us what is logically necessary (very little, but that little is
    crucial to all the rest). That which is not necessary, including
    the family and family enterprises, is analysed historically;
    however, insistence on its marginality to capital situates it in
    relation to what is theoretically central. The implications for
    possible change of such institutions or practices within
    capitalism are interesting and important.

4   For a different approach to the mechanical use of wage labour in
    defining capitalist production, see Mooney (1982). In
    agriculture, hired labour takes unusual forms. See the next
    section.

5   I include in formal exchanges the hiring of neighbours for wages.
    Such labour is available when operators are trying to expand in a
    phase of the enterprise cycle and when sons are not needed to
    work on the home farm in a phase of the family cycle (see Bennett
    1982:152, Friedmann 1978b).

6   Goodman and Redclift (in this volume) claim that it is populist
    ideology to distinguish family farms from 'small capitalist
    farms' subordinated to varying degrees by capital in the
    agricultural sector.

7   The social insurance dimension of deficiency payments, which are
    paid out of general state revenues and are targeted to incomes,
    might conceivably be enhanced and connected to both structural
    policies for agriculture and more general economic and social
    rights. Price supports, by contrast, apply to commodities and
    reinforce the property dimension of simple commodity production.

REFERENCES

Bennett, W. 1982. Of time and enterprise: North American family farm management in a context of resource marginality. Minneapolis: University of Minnesota Press.

Bombal, J. and Chalmin, P. 1980. L'Agro-alimentaire. Paris: Presses Universitaires de France.

Bogue, A. G. 1968. From prairie to cornbelt. Chicago: Quadrangle Books.

Bouquet, M. 1982. Production and reproduction of family farms in South-west England. Sociologia Ruralis 22, 227-244.

Briskin, L. 1985. Theorizing the capitalist family household system: a marxist feminist contribution PhD Thesis, York University, Toronto.

Burawoy, M. 1979. Manufacturing consent: changes in the labor process under monopoly capitalism. Chicago: University of Chicago Press.

Friedland, W., Barton, A. E. and Thomas, J. 1981. Manufacturing green gold: capital, labor, and technology in the lettuce industry. Cambridge and New York: Cambridge University Press.

Friedland, W. and Nelkin, D. 1971. Migrant: agricultural workers in America's Northeast. New York: Holt, Rinehart and Winston.

Friedmann, H. 1978a. World market, state and family farm: social bases of household production in the era of wage labour. Comp. Stud. Soc. Hist. 20, 545-586.

Friedmann, H. 1978b. Simple commodity production and wage labour in the American plains. J. Peas. Stud. 6, 71-100.

Friedmann, H. 1980. Household production and the national economy: concepts for the analysis of agrarian formations. J. Peas. Stud. 7, 158-184.

Friedmann, H. (forthcoming 1986) The family farm in advanced capitalism: outline of a theory of simple commodity production in agriculture. In The political economy of agriculture in advanced industrial societies, F. H. Buttel and T. Murphy (eds). New York: University Press of America.

Gasson, R. 1984. Farm women in Europe: their need for off-farm employment. Sociologia Ruralis 24, 216-228.

Goodman, D., Sorj, B. and Wilkinson, J. 1984. Agro-industry, state policy, and rural social structures: recent analyses of proletarianisation in Brazilian agriculture. In Proletarianisation in the third world, B. Munslow and H. Finch (eds), 189-215. London: Croom Helm.

Gussow, J. D. 1984. Food security in the United States: a nutritionist's view. In Food security in the United States, L. Busch and W. B. Lacey (eds), 207-230. Boulder and London: Westview Press.

Jackson, W. and Bender, M. 1984. An alternative to till agriculture as the dominant means of food production. In Food security in the United States, L. Busch and W. B. Lacey (eds), 27-45. Boulder and London: Westview Press.

Johnson, L. 1982. The seam allowance. Toronto: The Women's Press.

Kohl, S. B. 1976. Working together: women and family in Southwestern Saskatchewan. Toronto and Montreal: Holt, Rinehart, and Winston of Canada.

Laslett, P. 1965. The world we have lost. London: Methuen.

Lem, W. V. (forthcoming) Les rapports interpersonels de la production marchandise simple: les entreprises familiales à Murviel-lès-Béziers. In Les transmissions idéologiques et symboliques de la parenté en Occitanie, L. Assier-Andrieu (ed). Paris.

Mann, S. and Dickinson, J. 1978. Obstacles to the development of a capitalist agriculture. J. Peas. Stud. 5, 466-481.

Massey, D. and Catalano, A. 1978. Capital and land: landownership by capital in Great Britain. London: Edward Arnold.

Mintz, S. 1985. Sweetness and power: the place of sugar in modern history. New York: Viking.

Mooney, P. H. 1982. Labour time, production time, and capitalist development in agriculture: a reconsideration of the Mann-Dickinson thesis. Sociologia Ruralis 22, 279-292.

Mooney, P. H. 1985. The transformation of class relations in Wisconsin agriculture: 1945-1982. PhD Thesis, University of Wisconsin Madison.

Murray, R. 1978. Value and theory of rent: part two. Capital and Class 4, 11-33.

Newby, H. 1980. Green and pleasant land? Harmondsworth: Penguin.

Sabel, C. 1982. Work and politics: the division of labor in industry. Cambridge and New York: Cambridge University Press.

Scott, A. M. 1979. Who are the self-employed? In Casual work and poverty in third world cities, R. Bromley and C. Gerry (eds), 105-129. Chichester and New York: John Wiley and Sons.

Symes, D. G. and Marsden, T. K. 1983. Complementary roles and asymmetrical lives. Sociologia Ruralis 23, 229-241.

Thomas, R. J. 1981. Social organization of industrial agriculture. Insurgent Sociologist 10, 5-22.

Thomas, R. J. 1985. Citizenship, gender, and work: social organization of industrial agriculture. Berkeley and Los Angeles: University of California Press.

Walker, R. 1985. Is there a service economy? The changing capitalist division of labor. Science and Society 44, 42.83.

Zukin, S. 1982. Loft living: culture and capital in urban change. Baltimore and London: Johns Hopkins University Press.

4 The development of family farming in west Devon in the nineteenth century[1]

MICHAEL WINTER

INTRODUCTION

One of the characteristics of British rural sociology in recent years has been the attention devoted to family farming. In most sectors of British agriculture family labour has increased as a proportion of the total farm labour force throughout this century, and especially since 1945. This has caused something of a 'problem' for traditional sociological analysis. Marxist and, to a lesser extent, Weberian social theory confidently predicted the demise of family enterprises, which were seen as laggardly and traditional. Weber (1970), notwithstanding his recognition that certain conditions might favour small-scale farming within a rationalized market economy, generally expected each branch of economic activity to be characterised by an increasing level of bureaucracy and a complex division of labour. Marx, and more especially Lenin, envisaged a continuing process of differentiation in agriculture, with peasants emerging as fully-fledged capitalists or sinking into the army of the landless labourers. The idea has proved so beguiling that it can still be faithfully reproduced. Ernest Mandel (1978) talks of a 'massive conquest of agriculture by big capital' accelerating 'the social division of labour in agriculture'. Instead of an army of landless labourers and a hierarchical division of labour contemporary farming is characterised by a complex armoury of technical aids at the disposal of a greatly diminished number of farmers and even fewer farm workers.

Between 1960 and 1979, for example, the number of workers in UK agriculture fell by roughly half, that of farmers by one third (Britton et al 1980). Nearly three-quarters of UK farms now employ no full-time hired workers. One geographical area where family farming is particularly strong is the pastoral western half of Britain. In West Devon, for example, 80 per cent of the full-time farm labour force was provided by the farmer and family in 1979. The Holsworthy area of West Devon is one of the most difficult farming areas in Devon, outside Dartmoor and Exmoor, with very poor soils, and a wet climate, and has recently been accorded Less Favoured Area status. Although most of my work in the area has been on the contemporary structure of family farming, inevitably this has raised questions concerning the origins of family farming which I address here in an effort to unravel some of the competing general statements about historical processes of agricultural change in Britain.

SIMPLE COMMODITY PRODUCTION

Harriet Friedmann (1980, 1981) suggests a double specification of a simple commodity form of production, whereby factors, internal and external to the family unit, 'determine the conditions of reproduction of the form and the manner in which its circuits of reproduction intersect with those of other classes'. Friedmann's term 'simple commodity production' provides an important theoretical underpinning to this paper. However in the empirical sections of this paper 'family farming' is used in the 'everyday' sense to refer to producers primarily reliant on family labour. Similarly 'capitalist farmers' are defined empirically by their use of hired labour. The terms simple commodity production and family farming are not necessarily synonymous, as Friedmann's discussion of peasants and simple commodity producers indicates. She uses simple commodity production as 'a logical concept, referring to the complete separation of the household from all ties except those of the market' (1980:163). The use of the term 'family farming' allows the presentation of certain empirical positions prior to some theoretical conclusions.

Neither the internal dynamics of the unit of production nor the externally dominant mode of production are the sole determinants of the conditions of production and reproduction. Of crucial importance in Friedmann's analysis is the extent of commoditisation, meaning the degree to which family producers are integrated into the wider market economy for inputs and outputs. In addition, Friedmann attaches critical importance to the nature of that integration:

> Whatever the level of specialisation in production of commodities, if household reproduction is based on reciprocal ties, both horizontal and vertical, for renewal of means of production and subsistence, then reproduction resists commoditisation.
>
> (1980:163).

Such ties can include communal obligations to farming neighbours and kin, and non-market duties and obligations between landlords and tenants. Other forms of resistance to commoditisation will, of course, include subsistence production and restricted markets in land or labour. Widely varying degrees of resistance or openness to commoditisation enable us to distinguish between different types of family producer. It is to be expected that the nature of family farming will vary over time.

One of Friedmann's main research topics has been the rise of specialised household commodity production of wheat in Canada at the turn of the century which she likens to the 'new' family farmers emerging in England at the same time, especially in livestock production and dairying (Friedmann 1978a, 1978b). Both groups were highly commoditised in Friedmann's terms. Unlike 'peasants', defined by their resistance to commoditisation and often reproduced only through economic and political relations of dependence, simple commodity producers engage in free contractual relations within the capitalist economy. Therein lies the novelty of these specialised household producers. There is nothing in Friedmann's theory to suggest that simple commodity production should emerge at any particular point in the development of capitalism. Nevertheless she does point to the Great Depression at the end of the nineteenth century as a crucial period in the expansion of simple commodity production in England (Friedmann 1978a). Her emphasis on the

significance of the Great Depression is largely drawn from the
analyses of Fletcher (1960) and Perry (1974). A problem with these
works is that neither centrally addresses the issues of labour
relations and forms of production.

THE HISTORICAL EVIDENCE

Uncovering the evidence to determine the extent of simple commodity
production in particular areas during the nineteenth century is by
no means easy. In the introduction to his <u>Farmlife in Northeast
Scotland</u> Ian Carter (1979) traces the origins of his interest in
this topic to his desire to understand local cultural features of
the region. Rightly he felt compelled to seek their origins in the
locally dominant industry, agriculture. Carter recalls his early
searches in this direction:

> I began to look for a good modern local agrarian social
> history. I am still looking: it does not exist. Nor, to a large
> extent, can one find a corresponding economic history.
>
>   (p3).

Carter's response was to produce such a history himself, written
over nearly a decade and one of the most comprehensive regional
agrarian social histories to have been produced. The position with
regard to Devon is very similar, notwithstanding the attention which
the West Country has attracted in terms of anecdotal local histories
and a number of noteworthy studies of aspects of agrarian history
from social and economic historians. For the nineteenth century, in
particular there is a dearth of secondary material.

The primary material needs to be sifted very carefully. At one
level it is clearly the case that the incidence of hired workers in
the farm labour force was considerably greater in the nineteenth
century than it is today. For Devon as a whole there were three
times as many farm workers as farmers in the early part of the
nineteenth century. By the 1970s the position had exactly reversed,
with three quarters of Devon's full-time agricultural labour force
in the 1979 Agricultural Census comprising farmers and members of
the farm family.

Any suggestions that Devon, as a peripheral and pastoral
county, might have lagged behind the rest of England in the

development of capitalist social relations of production cannot be
borne out by the figures on labour force composition. On the
contrary in 1831 a higher proportion of Devon farmers employed hired
labour than in England as a whole (Table 1).

Table 1 Occupational status of males aged twenty and over employed
in agriculture, 1831 (percentages)

Occupational status	West Devon	Devonshire	England
farmer/employers	23.2	19.4	15.0
farmer/non-employers	9.0	7.0	10.7
labourers	67.7	73.6	74.3

Source: National Census

While the high incidence of hired labour in the total labour force
composition is the most striking feature of Table 1, the break-down
of occupiers into employers or non-employers is also of interest.
Thus 26 per cent of Devon farms employed no labour, or to be
strictly accurate no males over the age of twenty, compared to 42
per cent for England as a whole. In West Devon a slightly smaller
proportion of hired workers were employed than in England and the
rest of Devon, although the difference is not great. The proportion
of non-employing farmers is 28 per cent. Notwithstanding the overall
dominance of capitalist social relations of production these figures
indicate a significant family farming component in the agricultural
structure of the early part of the last century. Table 2 shows the
changing labour profile of Devon's agriculture over a 90 year
period, starting in 1831. The trends are very clear, despite one or
two apparent anomalies (for example the number of farmers in 1921 is
curiously high).While the number of farmers over the period held
more or less constant the number of farm workers fell from around
44,000 in 1861 to 19,000 in 1921. This shows clearly the trend which
has continued in the post-war period. For the country as a whole
Britton et al (1980) have identified two periods of a particularly
marked decline in the agricultural labour force, 1860-1890 and
1950-1970, the first in response to recession in agriculture, the

Table 2 Composition of the Devon farm labour force, 1831-1921

	1831	1861	1891	1921
farmers	12,684	11,325	9,636	12,803
farmers' relatives:				
male	-	5,221	3,805	4,538
female	-	4,609	-	913
total farmers & male relatives	-	16,546	13,441	17,341
farm bailifs	-	338	355	377
labourers & servants:				
male	35,311	41,461	25,743	19,080
female	-	2,356	667	-
ratio of male employees per 100 farmers & family	-	250	194	112

Source: National Census

second to expansion. This is one of the telling factors in the rise of family farming, namely that the response to contrasting economic conditions has been to shed labour.

During the last decades of the nineteenth century agricultural labour was shed as a response to deteriorating product prices on the world market. Arable acreages declined and cost-cutting through mechanisation had begun. At the same time industry continued to absorb surplus rural labour (Saville 1957, Orwin and Whetham 1964). But this process of labour shedding does not necessarily mean that family farmers as a proportion of farm occupiers have increased in numbers. As already shown a significant minority of family farmers existed in the nineteenth century. It is possible that this proportion could remain constant with the capitalist farms shedding labour. Mingay (1962) reckons on one man for fifty acres of grassland and one man for thirty acres of arable in the Victorian period. Thus a 300 acre pasture farm required a labour force of at

least six men. With three or even four employees in 1931 such a
holding would have remained within the capitalist category. However
a 100-150 acre grassland farm might well have made the transition
from capitalist to family relations. So to fully understand the
implications for the development of family farming we need to look
at changes in farm size structure. Unfortunately tracing such
changes is difficult, with repeated changes in the compilation of
national and agricultural censuses throughout the period.
Fortunately a good record of the relationship between farm size and
labour force is available from the 1851 national census which
reaffirms the tentative findings from the 1831 census.

Table 3 Distribution of agricultural holdings by farm size and
labour force size, Devon, 1851 (no. of holdings)

Farm size groups (acres)	Number of men in farm labour force					
	0	1-4	5-7	8-15	15+	Totals
under 20	1,015	219	1	0	0	1,235
20 - 50	920	976	8	3	1	1,908
50 - 100	730	1,835	86	10	3	2,664
100 - 150	227	1,465	251	39	5	1,987
150 - 250	89	968	455	167	7	1,686
250 +	32	226	222	250	91	821
no acreage stated	0	73	22	4	5	104
totals	3,013	5,762	1,045	473	112	10,405

Without doubt agrarian capitalism exhibited considerably more
diversity in nineteenth century England than is sometimes
acknowledged. Two-thirds of all farms were farmed with the help of
no more than two hired workers, and over three-quarters were less
than 150 acres in size. However, in terms of overall agricultural
production, family farms were far less significant, only 37 per cent
of land being in farms of less than 150 acres. The picture that
emerges is essentially similar to that described by Carter for
Northeast Scotland. Devon, in common it seems with most counties of
England, was dominated numerically by small-medium farms, dependent

on family labour supplemented by hired labour especially at certain
points in the family development cycle.

WEST DEVON AGRICULTURE

The nature of capitalist social relations on large farms at this
time has been well documented (Charlesworth 1980, Dunbabin 1963,
1968, Hobsbawm and Rude 1969, Howkins 1977, Newby 1977, Wells 1979),
but insufficient attention has been paid to the role of family
farmers. With such a high proportion of small farmers, some
dependent on selling some of their labour, there is clearly more to
the social relations of production than the division between farmer
and landless labourer. At times the contrasts amongst farmers must
have seemed almost as great as between farmers and labourers, a
situation with echoes today. This is apparent from a number of the
contemporary observers of the agricultural scene. By the middle of
the eighteenth century the reputation of Devon agricultural
production had slumped from its high point in seventeenth century
England, when Cromwell praised it as the best in England (Havinden
1969). The cleric, Dean Milles (1755) surveyed Devon agriculture in
the eighteenth century and found in Ashwater, one of the larger
parishes in West Devon, the cider to be 'rough and bad enough' and
the size, shape, colour and breed of cattle 'not worth remarking'.

Forty years later William Marshall (1796) found Devon farmers
to be 'ignorant and conservative' and closer to the labouring class
than to emergent agrarian capitalists. Many had risen:

> from servants of the lowest class; and having never had an
> opportunity of looking beyond the limits of the immediate
> neighbourhood of their birth and servitude, followed implicitly
> the paths of their masters. Their KNOWLEDGE is of course
> confined; and the SPIRIT of IMPROVEMENT deeply buried under an
> accumulation of custom and prejudice.
>
> (pp106-7,184; quoted in Chambers and Mingay 1966).

In other parts of the county Marshall had some praise for the
cultivation of temporary grasses, but not so for West Devon which
had 'no regular course of management; and it must remain in this
predicament until turnips and potatoes shall be introduced after
wheat or oats, as a fallow crop for barley and ley herbage'
(pp135-6). The situation had not improved by the middle of the

nineteenth century:

> The farming of Devon is at the present time inferior to that of
> most of the counties of England...the advantages of a genial
> climate and a fertile soil needing only the application of
> greater capital, industry and skill, encourage the hope that
> Devonshire will before long reinstate itself in the position it
> held in the sixteenth century, when it was an example of the
> best farming of the age.
>
> (Tanner 1849).

Caird (1852) detected the seeds of improvement in an emerging
capitalist class of farmers, at the same time noting the existence
of the smaller producers:

> There are two classes of farmers in the county, one consisting
> of men with small holdings, little elevated above the condition
> of the labourer, the other of educated agriculturists holding
> large farms into which they have introduced improved methods of
> husbandry. By them draining has been introduced, and the
> levelling of hedgerows and enlargement of arable fields; the
> system of irrigated meadows has been extended, and the
> application of artificial manures practiced. The improvement of
> the breed of Devon cattle, now one of the most shapely,
> graceful, and profitable breed in Great Britain, has been by
> them brought to its present high state of perfection.
>
> (p48).

Vancouver in 1808 similarly observed the contrast between the
smaller and larger farms and noticed a regional dimension. The
larger farms occupied the better lands of south and east Devon. In
the rich country south of Dartmoor farms with rental values of 5-700
shillings per annum were found. This contrasted sharply with West
Devon where the farms were mostly smaller. The West Devon District
is described in the following terms:

> Although there must necessarily be many exceptions to the
> general size of farms, and character of the occupiers, of a
> country so widely extended as this district, it may still not
> be amiss to go a little farther than ordinary into a
> description of these matters. In the country about East and
> West Putford, the size of the farms are not represented to
> exceed 20/- to 50/- per annum; the farmers, though a
> hard-working people, are supposed to remain stationary with
> regard to aquisition of property,.... About Holsworthy, the
> occupations are stated to be from 30/- to 50/- per annum; the
> farmers are equally laborious.... In the neighbourhood of
> Ashwater, the occupations are even smaller than about
> Holsworthy: these farmers are very industrious, working in
> general, much harder than the labourers they occasionally
> employ.
>
> (p108).

Later in the century, during the great depression, two Royal
Commissions on Agriculture provided further evidence on the
incidence of small-scale farming in Devon.

R. Henry Rew (1895) reported of North Devon that it was a
district of large estates and small farms. Rew also noted a tendency
for the number of small farms to increase during this period:

> There is some tendency on the part of landowners to divide
> farms where possible, and only the outlay necessitated in
> making such alterations probably prevents this tendency from
> being more marked. Where, as is not infrequently the case,
> farms were thrown together 20 or 30 years ago, opportunity is
> now taken when possible to re-divide them. There is greater
> demand for the smaller farms, and landowners are disposed to
> prefer the small working farmers.
>
> (p15).

Rew also comments on the incidence of small-holders engaged in
part-time agricultural work. He cites one parish in particular where
a landowner had encouraged the formation of such small-holdings.
Rider Haggard (1906), another seasoned observer of rural England,
cites evidence from the same parish, Beaworthy, where the number of
holdings is said to have increased from 243 in 1871 to 434 in 1901.

Overall a picture emerges of an agriculture adapting to the new
markets. Subsistence production is still important and clearly an
important factor in the survival strategies of smaller and larger
producers in these depressed years. But it is 'the admirable produce
markets in all the market towns' which bring forth particular
comment from Rew:

> It is the regular practice for the wives and daughters of the
> small farmers, or the dairywomen of the large farmers, to take
> poultry, eggs, butter, and clotted cream, as well as garden
> produce, honey, etc, into the market once a week.
>
> (p19).

All this evidence points to continuity and renewal of family farming
in the nineteenth century. This must act as something of a
corrective to the received wisdom of agrarian history concerning the
universality of capitalist agriculture in England from the eighteenth
century onwards. Mingay is one of the few agrarian historians to
have grasped the importance of family farming in the eighteenth and
nineteenth centuries issuing a challenge to those historians who

brush aside the evidence:

> It is evident, then, that in spite of the changes of the
> eighteenth and nineteenth centuries - in spite of the growth of
> the market, new techniques, enclosure, industrialisation,
> depression and free trade - the small farmers in general
> suffered no catastrophic decline in numbers. Anyone who is
> willing to believe that small farmers 'disappeared' in the
> eighteenth century must be prepared to explain how it was they
> re-appeared in such strength in the nineteenth century.
> (p470).

Mingay also highlights the importance of part-time farming,
estimating that any eighteenth or nineteenth century occupiers of
less than 25 acres would have needed an additional income. This
would have accounted for some 15 per cent of Devon farmers in the
1850s, possibly slightly more. These findings raise doubts about the
notion of a new specialist family farmer emerging in England at the
turn of the century as a response to new capitalist markets. Thus
for Friedmann (1978) small holders 'are not survivors in any sense
from the sixteenth century but a new kind of household producer, a
specialised commodity producer'. Markets for commodities such as
milk and poultry were expanding at the end of the century. The
market acumen of small Devon farmers has been testified to, but at
the same time this was combined with a quasi-subsistence (Perry
1974) orientation. Clearly family farming became increasingly
specialised and commoditised in the twentieth century - in Devon
more so in the post 1945 period than at the turn of the century
identified as crucial by Friedmann - and the proportionate
significance of family farming increased steadily. That much is well
recognised, but in the sense that small family holdings were
significant throughout the nineteenth century and earlier there is a
measure of continuity and survival.

Having discussed the labour composition of agriculture in West
Devon during the late nineteenth century and early twentieth
century, it is also important to give consideration to another
factor of key significance to the local development of family
farming. In common with the rest of England, Devon was made up of
primarily tenanted land in the nineteenth century. In 1900 just 15
per cent of land in the survey parishes, only a slightly higher
percentage than for Devon as a whole, was owner-occupied. However

this masked considerable variation between parishes. While a quarter
of parishes were clearly dominated by the landlord-tenant system
with less than 5 per cent of the land owner-occupied, for another
quarter the corresponding figure was over 20 per cent. The survival
of small ownership in Devon during the period of expropriation,
1660-1760, is revealed in the Land Tax assessments studies by
Johnson (1909, cf. Beckett 1984). A century later the famous 1873
'Domesday' Survey of landowners revealed Devon to have a higher than
average number of small landowners (Bateman 1873), either as small
landlords or owner-occupiers.

This incidence of small ownership led to a more rapid decline
of the landlord-tenant system than in other parts of England.
Owner-occupiers were inevitably in a stronger position than tenants
at the outset of the twentieth century. Economically the absence of
rent meant that owner-occupiers could survive in times of depression
and expand more rapidly when conditions improved. As one
owner-occupier told Henry Rew he 'would have gone to smash years
ago' if he had had to pay rent. Politically also they were freer
to express opposition to the power of the large landowners. Such
opposition in West Devon was strongly associated with the growth of
the Bible Christian movement, a Methodist sect primarily made up of
farmers and rural artisans. It is no coincidence that the movement
originated in the parish of Shebbear. As early as 1831 over half the
farmers of this parish employed no hired labour. In 1871 there were
no less than 46 resident owners of land in the parish, and in 1900
27 per cent of the land area was owner-occupied.

The growth of Methodism, allied to Liberal politics, proved a
potent political challenge to landlordism in contrast to the south
and east of England where past disagreements between landlord and
tenant were patched up in an alliance opposed to local organised
labour (never strongly developed in Devon). Rew, reporting to a
Conservative government, was obviously much disturbed by demands
from tenants for more freedom and greater security. Economically the
relative prosperity of farming in the west meant fewer unlet farms
during the depression years at the end of the century (Rew 1895).
Thus existing tenants achieved the security of tenure from new
legislation which later put them in a good position to purchase

their holdings from landlords weakened by low rents and high taxes.
In addition there is evidence that a large number of the smaller
landowners themselves turned to farming on their own account. By
1979 80 per cent of the land in the Survey Parishes was
owner-occupied compared to 60 per cent nationally (UK).

CONCLUSION

The preceding discussion raises a number of questions concerning the
nature of temporal change, not to mention interpretive problems
about continuity and disjuncture. Rapidly changing conditions of
production may throw up new forms of production in certain
circumstances, but permit adaptive survival strategies in another.
By the end of the nineteenth century a number of conditions combined
to increase the significance of family farming. Some family farmers
were 'new' in the literal sense of having arisen in response to new
markets, notably the rise of small-scale dairying (although dairying
was not nearly so well developed in Devon as in Somerset and
Cornwall). Others were 'new' only in the sense that they had shed
hired labour in the face of changed economic and technological
conditions. And some family farmers could be classified as 'old' in
the sense of representing a continuity of underlying form, the
'traditional' family farmers. Clearly some of the latter group
become increasingly marginalised but it is impossible to find a
sharp disjuncture between 'old' and 'new'. The new/old dichotomy
belies the essential structural unity of family production as a
response to changing conditions. To the family farmer of the
nineteenth century it made little difference whether he had recently
shed the final unit of hired labour or had always operated with
solely family labour. Indeed many family farmers were likely to have
alternated between family and hired labour according to the stage in
the family development cycle. If the new/old dichotomy presents
problems so too does that between capitalist and family producers,
especially if it is the capitalists who are identified with the
'older' order.

Finally what of the ideas raised in connection with Friedmann's
work at the outset of the paper? It is clear that the data for Devon
supports her assertions concerning the increase of small-scale

family farming during the depression years. However the evidence on the extent of commoditisation is only partial, not the least of the problems being insufficient evidence on the precise economic organisation of small farm businesses at the time. As one contemporary observer lamented 'few farmers keep accurate accounts ...small farmers never do' (Read 1887). The available evidence seems to point in different directions. It emphasises the frugal, quasi-subsistence existence of many small producers on the one hand. On the other hand the market for a number of products was expanding though the 'new' family producers had certain 'peasant-like' qualities. Perhaps the best way to conceptualise the change is as a convergence of two quite different groups each employing different survival strategies in the light of changing conditions of reproduction.

FOOTNOTE

1 In preparing this paper I am grateful for the comments of Peter Hamilton, Graham Cox and Philip Lowe. The paper is based on work carried out as part of a Ph. D. project at the Open University. The thesis is due for submission in Spring 1986. For some further indications of the line of research see Winter, M. (1983, 1984a, 1984b). A number of aspects of this paper are dealt with in more detail in the thesis. 'West Devon' in this paper refers to some forty parishes centred on the market town of Holsworthy in the north-west of the county. It includes a handful of parishes in Cornwall. Administratively most of the area falls into the Torridge District, rather than West Devon Borough which is further to the south centred on Tavistock and Okehampton.

REFERENCES

Bateman, J. 1873. The great landowners of Great Britain and Ireland. London.
Beckett, J. V. 1984. The pattern of landownership in England and Wales, 1660-1880. Econ. Hist. Rev. (2nd ser) 37, 1-22.
Britton, D. K., Burrell, A. M., Hill, B. and Ray, D. 1980. A statistical handbook of UK agriculture. Wye College.
Caird, J. 1852. English agriculture in 1850-51. Letter VII Devonshire, 48-56. London: Longman.
Carter, I. 1979. Farmlife in northeast Scotland 1840-1914. Edinburgh: John Donald.
Chambers, J.D. and Mingay, G. E. 1966. The agricultural revolution 1750-1880. London: Batsford.
Charlesworth, A. 1980. The development of the English rural proletariat and social protest, 1700-1850: a comment. J. Peas. Stud. 8, 101-111.
Dunbabin, J. P. D. 1963. The 'revolt of the field': the agricultural labourers' movement in the 1870s. Past and Present 26, 68-97.
Dunbabin, J. P. D. 1968. The incidence and organisation of agricultural trade unionism in the 1870s. Agric. Hist. Rev. 16, 114-141.
Fletcher, T. W. 1960. The great depression of English agriculture 1873-96. Econ. Hist. Rev. (2nd ser) 13, 417-432.
Friedmann, H. 1978a. World market, state and family farm: social basis of household production in the era of wage labour. Comp. Stud. Soc. Hist. 20, 545-586.
Friedmann, H. 1978b. Simple commodity production and wage labour in the American Plains. J. Peas. Stud. 6, 71-100.
Friedmann, H. 1980. Household production and the national economy: concepts for the analysis of agrarian formations. J. Peas. Stud. 7, 158-184.
Friedmann, H. 1981. The family farm in advanced capitalism: outline of a theory of simple commodity production in agriculture. Paper presented to the American Sociological Association, Toronto.
Haggard, Sir H. Rider. 1902. Rural England. London: Longmans.
Havinden, M. A. 1969. Agricultural history in the south west. In The south west and the land, M.A. Havinden and C.M. King (eds), Exeter Papers in Economic History, No 2.

76

Hobsbawm, E. and Rude, G. 1969. Captain Swing. London: Lawrence and
Wishart.
Howkins, A. 1977. Structural conflict and the farm-worker: Norfolk
1900-1920. J. Peas. Stud. 4, 217-229.
Johnson, A. H. 1909. The disappearance of the small landowner.
Reprinted (1963), Oxford: Oxford University Press.
Mandel, E. 1978. Late capitalism. London: Verso.
Marshall, W. 1796. The rural economy of the west of England.
Reprinted (1970) Newton Abbot: David and Charles.
Milles, Dean. c1755. Mss, Devonshire Mss, Bodleian Library, Oxford.
Mingay, G. E. 1962. The size of farms in the eighteenth century.
Econ. Hist. Rev. (2nd ser) 14, 469-488.
Newby, H. 1977. The deferential worker. London: Allen Lane.
Orwin, C. S. and Whetham, E. H. 1964. History of British agriculture
1846-1914. London: Longmans.
Perry, P. J. 1974. British farming in the great depression
1870-1914. Newton Abbot: David and Charles.
Read, C. S. 1887. Large and small holdings: a comparative view.
J. Royal Agric. Soc. 23, 1-27.
Rew, R. H. 1895. Report on north Devon, in Report of the Royal
Commission on Agricultural Depression, 1894-97, Parliamentary
Paper, XVI, (C.7728).
Saville, J. 1957. Rural de-population in England and Wales
1851-1951. London: Routledge and Kegan Paul.
Tanner, H. 1849. The farming of Devonshire. J. Royal Agric. Soc. 9.
Vancouver, C. 1808. A general view of the agriculture of the county
of Devon. Reprinted (1969) Newton Abbot: David and Charles.
Weber, M. 1970. Capitalism and rural society in Germany. In From Max
Weber: essays in sociology, H. H. Gerth and C. W. Mills (eds),
363-385. London: Routledge and Kegan Paul.
Wells, R. A. E. 1979. The development of the English rural
proletariat and social protest, 1700-1850: a comment. J. Peas.
Stud. 6, 115-139.
Winter, M. 1983. Family farming in west Devon: changing patterns of
social and economic organisation of pastoral agriculture. In
Strategies for family-worked farms in the UK, R. Tranter (ed),
191-195. CAS Paper 15, University of Reading.
Winter, M. 1984a. Corporatism and agriculture in the UK: the case of
the Milk Marketing Board. Sociologia Ruralis 24, 106-119.
Winter, M.1984b. Agrarian class structure and family farming. In
Locality and rurality: economy and society in rural regions,
T. Bradley, and P. Lowe (eds) 115-128. Norwich: Geobooks.

5 Part-time farming: its place in the structure of agriculture

RUTH GASSON

INTRODUCTION

A prevalent, though neglected form of agricultural organisation is part-time farming. The observations reported here are based on a pilot study of part-time farming families in England and Wales which was initiated by a group of agricultural economists from the universities of Oxford, Reading, Exeter and Wye College and carried out in 1981-2.[1]

'Part-time farming' is one of those loose phrases which slip easily off the tongue, whose meaning has become elastic by continuous stretching to accommodate different ideas. Alternatives like 'multiple job holding' 'farming with other gainful activities' and 'rural pluriactivity' are clumsy, harder to articulate and scarcely any more explicit. In this chapter I will stay with 'part-time farming' since it is the most familiar term and the one most often used in discussions of farm structure, agricultural policy and the rural economy. First I will consider alternative meanings of 'part-time farming' and explain the sense in which it is used here.

CONCEPTIONS OF PART-TIME FARMING

Part-time farming can be defined in terms of the farm, the time the farmer spends on the farm or the farmer's activities and income sources. Part-time holdings, in the sense of agricultural holdings not providing enough work to keep one able bodied person fully occupied, account for nearly half the census holdings in England and

Wales. This definition, based on the estimated labour requirement of the holding, is useful in certain contexts. It gives a fair indication of the problems likely to be encountered in such businesses: high fixed costs relative to output; problems due to the lumpiness of inputs, notably the farmer and wife's labour input; hence low labour productivity; difficulties in reaping economies of size; therefore low farm income. Partly following from these disadvantages but compounding them, little research, development of appropriate technology, advice or capital grants are directed towards part-time holdings.

The drawback with this approach is not knowing what role the part-time holding actually plays. It may be only a fragment of a larger farm business. It may be a sideline or one of several enterprises run by the occupier. It may not even be used for agriculture. It may, on the other hand, be a farmer's only activity; and his apparent under-employment may reflect a temporary stage of development, or an under capitalised business, or a farm in difficult terrain requiring more labour than the average. The farmer may have to support himself entirely from the farm or he may have other means. Therefore we need to know more about the relations between the holding and the occupier before pronouncing it 'a problem'.

Turning to definitions based on the activities of the occupier, between 20 and 25 per cent of all farmers, partners and directors on agricultural census holdings are part-timers in the sense of working less than full-time on the holding. Defining part-time farming in this way may be etymologically correct but is not particularly useful, unless we know more about the relationship between the farmer and the holding. The farmer or partner may be a part-timer because he or she has another job, devotes time to unpaid activities like housework or voluntary work, is physically incapable of full-time working or because there is not enough to do on the farm. Use of the farmer's time is not, in practice, a very objective measure. It is left to the discretion of the person completing the agricultural census form to decide what is full-time and part-time working. Our pilot study demonstrated how ambiguous the concept of working time can be. If, for example, an elderly farmer spends 35 hours a week pottering about on a small holding because he has nothing else to do while a younger man works hard for 40 hours a week on his farm and spends another 40 hours building up another business, which is the part-time farmer?

Defining part-time farming in the sense of the farmer having another gainful occupation is much more useful. The fact that the holding is not the only source of income has obvious implications for household consumption levels, for the family's financial security and attitude to risk, for farm investment and development, for the approach to the farm business as manifested by supply response, the drive to maximise profit, attitudes towards nature conservation and so on. The fact that the farmer divides his time between farming and another job has implications for labour productivity and for the time the farmer can spare for training and advice, and for farm planning and development.

Between 20 and 25 per cent of farmers in England and Wales are part-time in the sense of having other gainful activities. This is a low proportion compared with other advanced societies. In the European Community in 1975, 28 per cent of all occupiers had other work, in the United Kingdom only 21 per cent (Commission of the European Communities 1980). In the highly industrialised OECD countries in the late 1970s, 40 to 60 per cent of all farmers or farm households derived more than half their incomes from non-farm sources (OECD 1978). In 1978, 55 per cent of United States farm operators reported some off farm work (Buttel 1982) while over half the income of census farm operators in Canada comes from off farm work (Bollman 1979).

Potentially more useful still would be a definition based on the activities of the farm household, since decisions about income, investment and consumption, employment and leisure, apply to all household members. It would be helpful to have information on total farm household incomes and activities of all household members contributing income, whether or not they happened to work on the farm. This is the approach taken in this chapter. There are no statistics on the extent of multiple job holding farm households in this country. Although the Ministry of Agriculture collects data periodically on farmers and spouses who have other gainful activities besides working on the holding, i.e. dual job holders, they do not include cases where one or more household members works exclusively on and others exclusively off the farm.

While there is substantial overlap between these different definitions, it is by no means complete. Data collected by MAFF for the 1975 European Communities' Farm Structure Survey showed, for instance, that only half (51 per cent) of occupiers of part-time holdings in England and Wales worked part-time on their holdings, the rest claiming to work full-time. While 39 per cent of occupiers of part-time holdings had other gainful occupations, 61 per cent said they did not. Conversely, on full-time holdings, 11 per cent of operators worked only part-time and 11 per cent had other gainful activities (MAFF 1976).

One consequence of using the term 'part-time farming' with a range of meanings is that attitudes and policy stances appropriate to one tend to be transferred to another. The notion that a part-time farmer is a dilettante, a dabbler, one not seriously committed to his farming, tends to spill over into a dismissive attitude towards all farmers who are dual job holders. Official thinking about part-time farming seems to be permeated with conceptions of small scale, uneconomic units with no future, producing a low and inadequate income. Implicitly, the part-time farmer seems to be portrayed as a failed full-time farmer. If full-time farming represents the norm, then part-time farming is seen as an unstable, transitory state of affairs which is inherently undesirable. 'Solutions' to the part-time farming 'problem' involve helping individuals to attain the status of full-time farmers or quit farming altogether. Assistance to continue in part-time farming would only prolong the agony.

While it is possible to identify farming families to fit these stereotypes, they do not represent the whole or even the norm. Findings from our pilot study and elsewhere indicate that total incomes of part-time farming families tend to exceed those of their full-time farming counterparts. Part-time farmers are found on large, expanding farm businesses as well as on static or declining small units. They are more likely to be new entrants to farming than retreating from it. Part-time farming can be a stable and permanent status as well as a transitional phase, one that is positively sought rather than merely tolerated.

DATA COLLECTION

Information was collected from a sample of farm households who were visited during the summer of 1981. Because this was a pilot study, intended to explore the extremes of variation likely to be encountered in the main survey, it was not appropriate to draw a random sample. Hence the results are not necessarily representative. About 30 sampling points were chosen representing all the regions of England and Wales. The Ministry of Agriculture provided a list of holdings in each area where there was good reason to expect that the farmer or spouse would have other jobs in addition to farming. A total of 600 holdings was selected and information obtained from 427, a response rate of just 70 per cent. (For further information on the methodology and results, see Gasson 1983.)

THE NATURE OF OTHER OCCUPATIONS

From our initial sample of 427 holdings we identified 345 part-time farming households according to our definition, that is families running the farm as their own business with one or more members having another gainful activity. It was most often the farmer who had the other job. While 83 per cent of our farmers were dual job holders, only 17 per cent worked exclusively on the farm while other members worked elsewhere.

Activities thought to be typical of part-time farming such as farmhouse bed-and-breakfast, direct sales of farm produce and agricultural contracting, proved to be minority interests. While a total of 144 farmers and other household members was involved in farm based recreation and tourism, direct retailing and other enterprises based at home and a further 85 worked on other farms in some capacity, there were 393 persons, two-thirds of the total, who had off farm jobs. These off farm jobs were usually full-time,

whereas those based at home or on other farms were typically
part-time or seasonal activities.

Predominant among off farm occupations were other businesses.
Of the 134 farm family members in business, exactly half were
owners, directors or managers of firms large enough to employ
non-family labour and half ran small family firms or one man
businesses. Company directors, manufacturers and stockbrokers came
into the first category, shopkeepers, garage proprietors and
electrical contractors being typical of the second.

The next largest group comprised manual workers (111
individuals), including here both skilled trades (printer,
mechanic) and unskilled (factory worker, lorry driver, County
Council roadman). The sample also included 77 professional workers
and 71 in other white collar occupations. The criterion for a
profession was a training period of at least three years (as for
lawyer, teacher, nurse). 'Other white collar' jobs comprised
clerical, secretarial, technical, service and sales posts.

Table 1 underlines the importance of business occupations,
particularly for farmers and farmers' husbands in the sample.
Professions accounted for a third of the wives' off farm
occupations, followed by other white collar jobs. A quarter of the
farmers and three-quarters of the working sons' off farm jobs were
manual while a majority of farmers' daughters were in lower status
white collar jobs.

Only about a third of the farmers with off farm jobs held
relevant qualifications or had undergone formal training. The
situation was different for working wives, 46 per cent of whom were
qualified and trained for their off farm occupations. This ties in
with the higher proportion of wives in professional posts. Taking
all kinds of other gainful activities, farm based as well as off
farm, a majority of dual job farmers in the sample were found to be
employers or self employed in both occupations, only 35 per cent
being employees. This runs counter to the notion that part-time
farming represents a proletarianisation of the family farmer.
Britain may be unusual in this respect. The 1975 European
Communities' Farm Structure Survey showed that the proportion of

Table 1 Off farm occupation by position in household

Type of occupation	Farmer	Wife	Husband	Son	Daughter	All	Numbers
	per cent of off farm occupations						
business	46.7	17.2	51.4	5.4	3.6	34.2	134
profession	17.9	33.4	28.6	0.0	7.1	19.6	77
other white collar	10.4	27.2	2.8	18.9	67.9	18.0	71
manual	25.0	22.2	17.2	75.7	21.4	28.2	111
all occupations	100.0	100.0	100.0	100.0	100.0	100.0	
numbers	212	81	35	37	28		393

farmers with non-farm jobs who were self employed in those jobs was
much higher in the United Kingdom (45 per cent) than in the
Community as a whole (26 per cent) (Commission of the European
Communities 1980).

While 43 per cent of farmers in the sample regarded farming at
home as their major activity, 57 per cent put some other job first.
Only about a quarter of the farmers' spouses in the sample put farm
work as their main activity, slightly more putting an off farm job
first and more still (32 per cent) describing themselves as
housewives first and foremost. As far as income is concerned
two-thirds of the farmers or working couples in the sample earned
more from other sources than from the holding, the remaining one
third making as much or more from farming. These findings are very
similar to those from Alan Harrison's 1969 survey of farm businesses
in England (Harrison 1975).

DYNAMICS OF PART-TIME FARMING

A popular conception of the part-time farmer is one who has been
forced to seek other employment in the face of a low or declining
farm income - the 'failed full-time farmer'. In the sample, however,
fewer than half (48 per cent) of all farmers had begun their careers
in farming and later diversified into other activities. More than
half had entered farming later in life from some other career. In
Scotland, too, Wagstaff (1970) found that the other job had come
first, chronologically, for a majority of farmers with other jobs.

The image of the part-time farmer in a state of transition
between full-time farming and a non-farm job, gives rise to the idea
that part-time farming is an inherently unstable activity. Our data
did not support this very strongly, either. When classified
according to their first and present main occupation, the largest
group was found to consist of 150 farmers for whom farming had never
been the major activity. Next in importance were 111 farmers for
whom farming had always been the major, if not the only, gainful
activity. Smaller numbers fell into the two transitional categories;
87 'leavers' had switched from farming to some other main occupation
and 67 'entrants' had made farming their main occupation although it

was not so originally. Thus a minority of farmers in the sample (37 per cent) had changed the priority of farming in the course of their working lives.

Exploring the notion of part-time farmers being obliged to seek other work to supplement farm incomes, respondents were asked what activity they would prefer if they were completely free to choose. Over half (58 per cent) of those who answered the question said they would prefer to farm full-time while 36 per cent put dual job holding as their first choice, the remaining 6 per cent preferring to give up farming altogether.

INFLUENCE OF FARM SIZE

From this pilot survey the 'typical' part-time farmer emerges as someone who runs another business, most likely off the farm and not necessarily linked in any way with agriculture. Although it does not require qualifications, it will probably be a full-time job, yielding considerably more income than his farming activities and he will probably regard it as his major occupation. More likely than not, it preceded farming in his career.

Although it is tempting to describe 'typical' cases and dwell on central tendencies, one of the most striking features of part-time farming is its diversity. Almost any occupation can, it seems, be combined with farming. This is not to say that the pattern is completely random. One of the variables which proved to be effective in explaining and understanding the pattern was farm size.

There is a suggestion that part-time farming is more prevalent on very small and very large farms than on medium sized family farms. Alan Harrison drew attention to this U-shaped distribution. While part-time farming predominated on farms below 50 acres in his national sample and full-time farming in the range 50 to 500 acres, he found that farms over 500 acres accounted for 4 per cent of all full-time but 6 per cent of all part-time farms. The 1980 Farm Structure Survey suggests a similar trend. On holdings above 2000 standard man days (roughly an 8 man farm), the farmer was twice as likely to have another gainful activity as on farms in the 1000 to 1500 smd range.

Table 2 Characteristics of the farmer's occupation by farm size

Farm size in smds	Farmer has an OGA*	Farmer alone has an OGA	OGA has ever been the main activity	Main activity now is farming	N.
		per cent of holdings			
under 50	77.5	41.7	85.0	27.5	119
50 - 99	65.6	28.9	77.8	35.6	88
100 - 249	51.9	26.0	58.7	59.6	101
250 - 499	75.6	40.0	68.9	44.4	45
500 and over	76.5	47.1	67.6	47.1	34
all farmers	67.8	34.9	73.0	41.5	387+

*OGA - Other Gainful Activity
+Total is less than 427 because smd ratings were not available for all holdings in the sample.

When other characteristics of part-time farming are analysed according to farm size, the U-shaped pattern stands out more distinctly. The size criterion used here is standard man days. A holding with a standard labour requirement below 100 smds could be run as a spare time activity with another full-time job. The theoretical dividing line between part-time and full-time working on the farm comes at 250 smds. Holdings above 500 smds theorectically require more than two full-time workers and are assumed to be commercially viable.

Farm families in the middle band stood out as being more 'farm centred' than those at either end of the size spectrum. Table 2 shows that middle band farmers in the sample were less likely than those at either end to have other occupations and much less likely to be the only household member with another gainful occupation. They were the least likely to have off farm jobs now and the least likely ever to have had a non-farm job as their major occupation. They were more likely than the rest of the sample to have begun their working careers in agriculture and the most likely to regard farming at home as the main activity.

Occupational status and earnings follow a similar trend. As farm size increased from 50 smds upwards, so did the likelihood that the farmer would be an employer or self employed in his other job rather than an employee. The larger the farm, the less likely it was that the farmer would be a routine white collar or manual worker. But among the very smallest farms, below 50 smds, there was also a concentration of farmers in high status business and professional occupations, giving a U-shaped distribution. Following from this, highest non-farm incomes tended to go with the largest and smallest farms in the sample. Predictably, farm incomes rose in line with the size of farm. Thus highest total earnings tended to be concentrated at the ends of the size spectrum, lowest in the middle. As Table 3 shows, middle band farmers, especially in the 100 to 250 smd range, were unusually reliant on farm incomes in the composition of total earnings, but these farm incomes were generally low.

Other studies have produced similar results. In the United States, for instance, farm families in the middle band have much

Table 3 Farm size and structure of farmer and spouse's earned income

Farm size in smds	Farm income over £5,000	Other earnings over £10,000	Farm is main income source
		per cent of holdings	
under 50	1.8	28.0	16.4
50 - 99	7.1	17.7	26.7
100 - 249	12.1	20.7	45.6
250 - 499	11.7	31.0	35.6
500 and over	59.3	33.3	62.1
all holdings	11.5	24.4	32.3

lower off farm earnings than those on either larger or smaller farms (Buttel 1982). From studies of part-time farming in a number of advanced countries, the OECD detected a general pattern:

> In areas where off farm employment is available a large percentage of farmers with non-viable holdings receive earned income from other sources: this income is comparable to that earned by normal wage earners in the district and is the dominant source of income received by the farm household. With increasing farm size, until the viable family farm size is reached, it is found that the percentage of farmers with off farm work declines, as also does the average income from this off farm work. Off farm employment, as well as income from this employment, then tend to rise slowly as the size groups above the family farm are reached.
>
> (OECD 1978).

In the light of these findings, it was not surprising that middle band farmers were much more likely to choose full-time farming as their preferred occupation and to reject multiple job holding, which appeared to bring them few rewards. As Table 4 shows, those on very small and particularly those on large farms, were more favourably disposed towards part-time farming.

Part-time farming on large, small and middle sized farms could have different meanings for the families involved. On very small holdings, capable of being run by the family at weekends, where the household is wholly or mainly supported by another source of income, living on a farm may be an expression of personal choice. It may be valued as an investment or a nest egg, as a place to live, for rural

Table 4 Farmer's preferred activity by farm size

| Farm size in smds | Preferred activity of farmer | | | |
| | full-time farming | part-time farming | other/no reply | Total |
	per cent of farmers			
under 50	44.2	31.7	24.1	100.0
50 - 99	52.2	25.6	22.2	100.0
100 - 249	60.6	20.2	19.2	100.0
250 - 499	44.4	42.2	13.4	100.0
500 and over	32.4	55.9	11.7	100.0
all farmers	49.4	30.5	20.1	100.0

surroundings, peace and quiet, for the lifestyle, purposeful
activity or as a focus for family enterprise. At the other extreme,
on holdings which are large enough to absorb all the family's labour
and more, and provide an adequate income without resort to other
employment, combining other activities with farming may also be an
expression of choice. A new activity may make use of spare resources
in the farm business, it may be a challenge, a means of contact with
a wider circle of people, a fulfilment of career ambitions. It is in
the middle band where families may have little choice but find other
activities to supplement their farming, that families are liable to
be disadvantaged and part-time farming resented.

SOME WIDER IMPLICATIONS

Part-time farming, involving about a quarter of the nation's farming
families and farm businesses, has wider implications. Criticisms of
this form of agricultural organisation tend to focus on economic
weaknesses of high cost production in small units, leading to wasted
resources and low efficiency. These are criticisms of small scale
farming rather than of part-time farming per se. As suggested
earlier in this paper, attitudes associated with part-time farming
in the sense of farming on non-viable holdings tend to be applied to
part-time farming as practised by multiple job households, even
though the latter may farm on a large scale.

In mitigation, it can be pointed out that the resources which
may be used wastefully by part-time farming families may have a low
opportunity cost. The principal resource is likely to be labour,
supplied mainly if not entirely by the family, for which the
alternative may well be leisure. Furthermore, while small scale
farming may constitute a farm income problem, existence of other
gainful occupations may overcome that problem.

Part-time farming has also been criticised for causing resource
immobility. Cushioned by their other income sources, part-time
farmers may be slow to respond to signals from the market. Because
they do not have to rely on farming for their living, they may hold
on to their farms in spite of low or negative returns, which may
prevent efficient, dynamic, bona fide full-time farmers from
enlarging their operations.

Leaving aside the fact that among the dynamic, expanding
farmers will be some part-timers, one can ask whether a degree of
resource immobility in the structure of agriculture would be
altogether undesirable. When over production poses more of a threat
to Britain than food shortages and when increasing value is attached
to conserving a pleasing and harmonious rural environment, part-time
farming may have some positive advantages. It helps to maintain a
network of small, varied farming units with their associated
infrastructure of dwellings, farm buildings, fields, trees and other
landscape features where the alternative, farm amalgamation, is
likely to destroy those features. Farmers who are not wholly
dependent on farming for their living can afford, if they so wish,
not to pursue maximum production at the cost of the natural
environment. In our sample, there was a highly significant
association between entering farming from some other occupation and
expressing a highly favourable attitude towards nature conservation
on the farm. Those who were mainly dependent on other income
sources, like the newcomers, were significantly more likely than
those wholly or mainly dependent on farm incomes, to have taken some
positive action to conserve the natural environment on their farms
and to be planning such action in the next two years.

The social impact of part-time farming may outweigh its purely
economic effects. An influx of families with different backgrounds,
occupations, lifestyles, expectations and values from the full-time
farming families they replace, could be a source of instability and
social conflict in rural communities. It could also have a
stabilising influence. Besides maintaining the number of farming
families, part-time farming may slow down the exodus of hired
workers. Part-time farmers in the higher status, better paid
occupations like professions and business, tend to employ workers
while they themselves are engaged off the farm. Loss of jobs on
farms is often associated with farm enlargement, which was not
proceeding at all rapidly among the part-timers in our sample. Since
a majority of part-time farmers run other businesses, there is a
chance of new jobs being created in the rural area. Taken together,
these tendencies may add up to greater stability for rural
employment and population and a slower rate of change in the
appearance of the countryside.

FOOTNOTE

1 Currently my colleague Berkeley Hill and I are conducting the
   main survey with a larger and more representative sample. Field
   work was completed in 1984 and final results should be available
   in 1986.

REFERENCES

Bollman, R. D. 1979. Off-farm work by farmers. Ottawa: Census
   Analytic Study, Statistics Canada.
Buttel, F. H. 1982. The political economy of part-time farming.
   Geo Journal 6, 293-300.
Commission of the European Communities, 1980. The agricultural
   situation in the community: 1979 report. Brussels/Luxembourg:
   The Commission.
Gasson, R. 1983. Gainful occupations of farm families. Wye, Kent:
   Wye College, School of Rural Economics.
Harrison, A. 1975. Farmers and farm businesses in England. Reading:
   University of Reading, Department of Agricultural Economics and
   Management, Miscellaneous Studies No. 62.
Ministry of Agriculture, Fisheries and Food, 1976. EEC survey on the
   structure of agricultural holdings, 1975: England and Wales.
   Government Statistical Service (Press report).
Organisation for Economic Cooperation and Development, 1978.
   Part-time farming in OECD countries: general report. Paris: OECD.
Wagstaff, H. R. 1970. Scotland's farm occupiers. Scot. Agric. Econ.
   20, 277-285.

6 Small scale farming in the Northern Ireland rural economy

JOAN MOSS

INTRODUCTION

The small, owner-occupied, family-run farm predominates in the
agricultural sector in Northern Ireland. This distinctive farm
structure is mainly the result of land legislation, some dating from
the end of the nineteenth century. Ownership of the land was
transferred to the sitting tenants and the absentee landlord system
was abolished. This transfer of ownership created a large number of
small farm units. The subsequent emphasis on fairly intensive
livestock enterprises generated sufficient income to ensure the
survival of numerous small farms. The land legislation also
curtailed agricultural tenancies, though 20 per cent of land devoted
to crops and grass is still let each year on a seasonal basis under
the conacre system. Employment in farming accounts for 10 per cent
of total employment in the region, with a further 3 per cent engaged
in ancillary jobs. Farming therefore contributes more to overall
economic activity in Northern Ireland than in the United Kingdom as
a whole, where less than 3 per cent of total civilian manpower is
employed on farms.

STRUCTURE OF FARMING IN NORTHERN IRELAND

Statistics on farm structure for the United Kingdom (including
Northern Ireland) are expressed in standardised labour units, with
farms in excess of 250 standard man days defined as full-time. The
statistics on farm structure published separately for Northern

Ireland are expressed in terms of standardised gross margins (total value of production minus variable costs) with 1 European Size Unit (ESU) = 1000 units of account of standard gross margin. Farms in excess of 4 ESU are defined as full-time. According to these separate definitions, in 1983 full-time farms in the United Kingdom averaged 119 ha of crops and grass, whereas in Northern Ireland, farms considered to be full-time were smaller and averaged only 42 ha. Even this comparison is misleading as it omits almost half of the 25,000 farm businesses enumerated in the June 1983 Northern Ireland agricultural census. These omitted 'small scale' farm businesses (of 1-3.9 ESU) averaged only 17 ha of crops and grass but they accounted for 33 per cent of Northern Ireland's full-time farm labour and generated 14 per cent of total standard gross margin.[1] This chapter will address their past performance and future role in the Northern Ireland rural economy with particular emphasis on those farmers who combine farming with off-farm employment.

The small scale farms, in common with their larger neighbours, have mainly livestock enterprises. Over 70 per cent concentrate on beef cattle and/or sheep production and account for 32 per cent of all slaughter cattle, 40 per cent of suckler cows and 28 per cent of all sheep in Northern Ireland. This has not always been the case. In former years dairying and pig and poultry production also featured on many small scale farms. A longitudinal analysis (Moss 1980) of small scale farms which compared level of business activity and areas of land owned and rented over the ten year period 1968-1977, showed that the majority of changes in farm business size measured in standard labour requirements resulted from switching from the more labour intensive enterprises of dairy, pig and poultry production to the less labour intensive production of beef cattle and sheep. Relatively little change occurred in the areas of land either owned or rented; hence disposals or acquisitions of land were not responsible for the vast majority of observed shifts in farm business size.

A survey of 200 beef cattle and sheep farms in the 100-199 standard man-day (at 1968 values) size range provided information on the operation and continued survival of these small scale farm businesses.

FARM LABOUR FORCE

On the majority of small scale farms surveyed the farmers were the
sole source of labour. Only 5 per cent of them were assisted
full-time by resident male family members and 8 per cent by resident
female family members. Part-time help from resident family members
was more common. Over 20 per cent of farms had part-time assistance
from male members and a similar proportion from resident female
members, usually the farmer's wife. Members of the farmer's family
not living on the farm can also be a useful source of farm labour.
No female help of this sort was recorded but 9 per cent of farmers
interviewed received part-time help from male relatives, usually
either brothers or sons. The majority of part-time work was supplied
for a couple of weeks each year during the demand periods of hay and
silage cutting and the remaining part-time labour was used for a few
hours each week, throughout the year, to assist with livestock
feeding.

Due to their relatively low labour requirements, beef cattle
and sheep enterprises are more compatible with off-farm employment
of the farmer than other enterprises such as dairying and cereal
production. Linear programming models indicated that the farms could
be run efficiently with less than 50 per cent of the potentially
available farm labour (generally the farmer's own labour). The
viability of a small scale farming system, however, may depend upon
the availability of contractors or extra hired labour during the
periods of peak labour demand. The lack of additional seasonal
labour may force farmers to engage in enterprises with fairly
uniform labour requirements throughout the year such as out-wintered
suckler cows or summer fattening of purchased store cattle.

FARM AND OFF-FARM SOURCES OF INCOME

Although the stocking rate on the small scale farms was estimated to
be equivalent to that found on larger beef and sheep farms, the
amount of farm income which could be generated was very modest,
given the limited land base. The low farm incomes also represented a
very low return on capital: in part this is a reflection of the high

land values in Northern Ireland which are such that the average capital value of the surveyed farms was over £100,000.

Forty five per cent of the farmers interviewed obtained more than half their total income from off-farm sources, but a further 40 per cent received very little off-farm income. Only 3 per cent received an investment income. Any spare capital was more likely to be invested in the farm business than elsewhere. Thirty per cent of the farmers were obtaining child allowances and some of these were also in receipt of a state retirement pension which reflected the late age at which many had married. In all 22 per cent of the small scale farmers received the old age pension but no privately financed or occupational pensions were recorded. A further 10 per cent of farmers received a family income supplement from the state. The importance of these transfer payments must not be underestimated. There is no doubt that the state pension, in particular, constituted a significant proportion of the total income of many of the elderly farmers.

More than half the farmers below retirement age engaged in off-farm employment. Over 80 per cent of these genuine part-time farmers estimated that more than half their total income was obtained from off-farm sources. Members of the farmers' families, living on the home farm but employed off the farm, also augmented family income. On a quarter of farms, at least one male family member (excluding the farmer) had off-farm employment; and on 13 per cent, a female family member (usually the farmer's wife) had off-farm employment. It was difficult to assess to what extent these off-farm incomes helped to finance the farm business. In the case of working wives, the contribution to the funds of the farm household may have been highly significant. Even if not used directly to finance the farm business, such income could ease domestic demands on farm income.

Few of the farmers surveyed were aiming for or achieving self-sufficiency in food: whereas one third of them consumed home grown potatoes, only 28 per cent had their own supply of eggs and less than 20 per cent their own vegetables or milk. Family

consumption of farm produce, therefore, did not substantially
supplement monetary household income.

PART-TIME FARMING

The desire to achieve an acceptable standard of living for their
families was cited as the motivating force by those who combined
off-farm employment with running their farms. This may explain the
significantly higher incidence of married men among the part-time
farmers. The off-farm jobs were neither seasonal (nor casual) in
nature nor part-time as has been frequently observed in other
countries (OECD 1978). On average they accounted for 37 hours per
week, 48 weeks of the year. When added to the average of 25 hours of
farm work per week for the part-time farms, it is evident that these
farmers had a substantial work load. They were certainly not hobby
farmers. Indeed, there was no significant difference between the
part-time farmers and those with no off-farm employment in the areas
of their farms. Not surprisingly though, the former were, on
average, 10 years younger than the latter (43 years compared to 53
years).

The off-farm jobs held included all aspects of employment
available in rural areas. The most frequently cited type of work (27
per cent of off-farm jobs) was with the building industry and ranged
from self-employed building contractors through tradesmen, joiners
and plasterers to unskilled labourers and employees in building
supplies firms. The second most common type of work (17 per cent of
jobs) was associated with transport, again ranging from
self-employed haulage contractors to bus and lorry drivers and their
assistants. Employment directly relating to agriculture such as
agricultural contracting and farm labouring, or ancillary to
agriculture, for example milk recording and cattle dealing accounted
for a further 16 per cent of off-farm jobs. The remaining jobs
varied in status and training from the veterinary surgeon, teachers,
nurses, policeman and clerks, to unskilled factory workers.

The three areas of employment often mooted as suitable for
part-time farmers - i.e. forestry, tourism and crafts - were not
well represented. Of the 82 farmers with off-farm employment only

three worked in forestry, two were associated with tourism (one running a caravan park on his land, another selling caravans), and none was involved in craft work.

Many farmers had obtained off-farm employment some time after inheriting their farms, having worked since childhood solely on the family farm. They explained that they had been forced to seek additional income to augment their farm incomes. The majority were engaged in either unskilled work or work where farming skills, such as driving, animal husbandry or basic building skills, could be put to good effect. Others had been employed outside of farming prior to succession and had decided not to relinquish their off-farm jobs when they inherited their farms. These farmers were more likely to have skilled jobs. Another group of farmers engaged in self-employed work such as contracting. This type of work has a long tradition in the farming community as a means of augmenting farm incomes, particularly for farmers' sons with an aptitude for mechanical equipment. The contracting work rarely ceases when the son inherits the family farm.

It has been said that there is a reluctance on the part of farmers to seek work off the farm because of lack of skills (Hathaway 1963). There was no evidence of this in the survey except among the more elderly farmers. While small scale farmers rarely possess formal qualifications they have a wide range of job skills due to the diversity of tasks which must be performed on any farm, irrespective of size. Indeed, from the evidence of the survey, farm location would appear to be a more important constraint on whether or not a farmer can obtain off-farm work, rather than possession of work skills: proportionately more of the farmers below retirement age with no off-farm employment lived in the remoter hill areas where alternative opportunities are limited.

FUTURE OF PART-TIME FARMING

Part-time farming is sometimes viewed as a transitional stage with farmers combining farming and off-farm work as a temporary measure prior to either leaving the agricultural sector or becoming full-time farmers. However, neither quitting farming nor progressing to full-time farming was cited as a prospect by the majority of part-time farmers interviewed. Only two had plans to sell land in the following 10 years and give up farming. Twelve farmers, representing 15 per cent of the part-time farmers had aspirations to become full-time farmers. Only some of these, however, had definite plans to buy land and expand their farm businesses in the following 10 years. The others were hampered by the lack of suitable land for purchase in the vicinity of their farms and shortage of capital to finance land purchase and other farm developments.

In contrast to the small proportion who had firm plans to buy additional land and become full-time farmers, over 60 per cent of the part-time farmers were hoping to make farm improvements in the near future ranging from drainage and re-seeding to building livestock housing. Most of these were willing to borrow to finance the farm developments but would not borrow to finance land purchase. The part-time farmers exhibited a greater willingness to borrow than the full-time farmers which probably reflected confidence in their ability to repay debt from their two sources of income. The commitment to the future of the farm businesses extended to the farmers' plans for succession. Over 70 per cent of all the small scale farmers interviewed believed that their farm businesses would remain as separate units after inheritance by their successors. Only 11 per cent claimed not to have an heir and these included a number of young unmarried men who would probably have heirs in the future.

POLICY MEASURES TO ASSIST SMALL SCALE FARMERS

Small scale farmers constitute a significant proportion of the agricultural labour force and hence they play an important role in the rural economy of Northern Ireland. If economic forces squeeze them out of farming, then rural communities would be diminished and unemployment increased. As they have a strong commitment to farming, assistance is needed, not so much to keep them in farming in the short term, but to alleviate their generally low incomes. This could be achieved by two mutually supportive strategies: first increasing farm incomes and secondly promoting opportunities for farm employment.

An economic analysis of the constraints on the growth possibilities of these small scale farms using linear programming techniques (Moss 1983) indicated that farm incomes could be raised if livestock housing was expanded. The small area of land per farm, however, imposes the ultimate constraint on the farm income which can be generated. Agricultural policies designed to boost farm incomes via product price support have had limited impact on small scale farms with their modest output.

The increase in livestock prices necessary to bring the small scale, full time, beef cattle and sheep farmers up to a standard of income comparable to that in non-farm sectors would impose a politically unacceptable cost on national or European Community budgets. Most livestock products are currently in surplus and any rise in consumer prices would exacerbate the problem by encouraging consumption to fall and output to rise. Furthermore, the main beneficiaries of higher product prices would be the large scale producers.

Many other forms of support specifically exclude small or part-time farmers. The Suckler Cow Premium, for example, was limited to farmers who obtained more than half their total income from farming. Such a restriction has, in the past, applied to many grants and development schemes together with the stricture that a certain target farm income had to be attainable. These requirements were intended to avoid uneconomic investments, even though on a number of physical measures of efficiency, such as stocking rate, small farms compare well with larger farm businesses.

Many small scale farmers have benefitted from the Agricultural Development Programme (ADP) which currently provides grant aid to farmers in the Less Favoured Areas whose farm businesses are incapable of attaining, at the completion of a farm plan, the comparable income per labour unit required for eligibility for the Agriculture and Horticulture Development Scheme. There is still the restriction, however, that at least half the farmer's total income must come from farming and more than half his labour must be devoted to the farm business. Moreover, at the end of 1984 the ADP grant rates were substantially reduced and ceilings were placed on the maximum amount of grant which could be obtained over a given period of time.

Even when small farmers do not fall foul of grant eligibility rules, they have less flexibility in their choice of enterprises than the larger farmers because of their limited scale of operations and labour supply. This inflexibility constrains their response to changes in the relative profitability of alternative enterprises and to a range of policies designed to influence farming systems, farm structures or farm income. In other respects, part-time farmers may enjoy more flexibility than full-time farmers, in being able to adjust the balance of the time they spend on and off the farm in response to the fluctuating profitability of farming, their changing capacity for work and its availability, and the changing income needs of their families. Policies designed to improve farm structures by financial inducements such as retirement annuities, payments to outgoers and farm amalgamation schemes have consequently met with a poor response from small scale farmers, particularly in times of economic recession. A low farm income may be preferable to no job at all especially if quitting farming entails relinquishing the family home.

Despite the low income earning potential of the numerous small farms in Northern Ireland, they do provide a source of employment in the rural areas, albeit part-time employment which needs to be augmented with off-farm work if satisfactory levels of income are to be achieved. The current small scale farmers undoubtedly have a long term commitment to their land and the rural economy. However, difficulties in obtaining off-farm employment may in the long term diminish their numbers. Potential successors are thereby inhibited from assuming responsibility for the family farm. This would have repercussions on the size of rural populations and the viability of rural communities. This need for off-farm employment must be viewed in the context of the recent history of unemployment in Northern Ireland. In the past 10 years the number of people registered as unemployed has risen from approximately 27,000 to 121,000, i.e. from 5 to 21 per cent of the workforce. This increase in unemployment has been uniform throughout the province's rural and urban areas and has not been concentrated in the two major conurbations of Belfast and

Londonderry. The lack of alternative employment has been a major short term obstacle for policies designed to improve farm structure and disperse surplus farm labour, particularly family labour.

Farmers, of course, are entitled to the full range of social welfare payments provided by the State. Recent debate on low farm incomes in the European Community, however, has focused on special direct income supplements analagous to the 'farmers' dole' in the Republic of Ireland. These could be an alternative to price policy as a means of providing sufficient income for small farmers who, due to age or remoteness, were unable to obtain off-farm work. Such a shift in policy, though, would encounter many objections. In particular, with millions of people unemployed in Europe and the vast majority of them lacking the capital assets of the typical small farmer, it would be difficult to justify guaranteeing the small farmer a standard of living above the social welfare threshold.

The proposed new Common Agricultural Structures Policy (Commission of the European Communities 1984) envisages an improved system of investment aids for small scale farmers. In response to increasing surpluses, reduced growth in agricultural incomes and the crucial problem of unemployment, it departs from the objective of raising farm incomes by increasing farm production irrespective of market conditions. A basic objective of the new structures policy will be the protection of employment in agriculture with an emphasis on assisting farmers in lower income brackets who have the greatest need of investment aid. It has been proposed that their incomes, living and working conditions should be improved mainly by a reduction in production costs, with specific assistance also available for appropriately trained young entrants to farming. The scope for reductions in production costs is fairly limited in the beef cattle and sheep systems pursued by the majority of the small scale farmers but preferential grants available to young entrants (under 40 years) would ease some of the longer term problems faced by successors wishing to assume control of family farms.

CONCLUSION

Most small scale farmers in Northern Ireland are determined to
retain family ownership of their land. With little prospect of
improvement in the levels of income generated on the small farms,
part-time farming will continue to be a permanent feature of small
scale agriculture. The extent to which off-farm employment can
augment farm income will be curtailed by the prevailing high levels
of unemployment. If policy makers consider it desirable to maintain
the size of the existing rural population, then policy measures
which discriminate against the smaller farmers will have to be
reassessed. Any attempt at introducing an integrated rural
development programme should take account of the job needs of those
requiring employment in addition to their farm work as well as of
those who are wholly unemployed.

FOOTNOTE

1 The 33% figure is as recorded in the Agricultural Census. It refers to the farm operator's own assessment of his/her input.

REFERENCES

Commission of the European Communities 1984. The new common agricultural structures policy. Brussels: The Commission.

Hathaway, D. E. 1963. Government and agriculture. London: Collier-Macmillan.

Moss, J. E. 1980. Part-time farming in Northern Ireland. A study of small scale beef and sheep farms. Studies in Agricultural Economics, Economics and Statistics Division, Department of Agriculture for Northern Ireland.

Moss, J. E. 1983. Growth possibilities for small scale beef and sheep farms. Studies in Agricultural Economics, Economics and Statistics Division, Department of Agriculture for Northern Ireland.

OECD 1978. Part-time farming in OECD countries. Paris: OECD.

7 Landownership relations and the development of modern British agriculture

SARAH WHATMORE

INTRODUCTION

The lack of theoretically informed analyses of the role of
landownership relations in modern British agriculture constitutes a
remarkable gap in contemporary agrarian and rural research. While
both agriculture and landownership have been marginalised within the
national political economy, land and the property rights governing
its use have remained central to the productive relations of
agriculture itself. This blindspot in agrarian research is all the
more remarkable because we have at our disposal the valuable
theoretical legacy of the nineteenth century classical economists,
most notably Marx and Ricardo, whose 'theories of rent' specifically
examined the ways in which the productive relations of agriculture
were modified by the ownership of land under a system of exclusive
property rights.[1] This body of theory has been revitalised in recent
years, however, not by those interested in contemporary agricultural
questions, but in the context of the urban land development process
and the production of the built environment (Ball et al 1984).

The lack of attention paid to this key component of the
development process and the structure of capitalist agriculture is
in part the result of the way in which agriculture has been studied
in the British academic tradition. In the first place, the
development of a political economy of agriculture (or indeed any
holistic analysis of agricultural production and social relations)
has been hampered by the separatism of 'rural sociology' and

'agricultural economics' as academic disciplines (Newby 1982b). The
social and economic aspects of agricultural production have remained
poorly related, limiting the explanatory value of the analyses
produced. This conceptual weakness has been compounded by the
broadly empiricist tradition which has characterised the
methodologies of both disciplines. In rural sociology the analysis
of agricultural relations has been largely subsumed within a widely
adopted, but ill-defined 'rural problematic' and an anthropological
or community-based social method. One of the most serious
consequences of such an approach has been the failure to locate the
productive relations of modern agriculture within the wider
structure of British capitalism. For example, contemporary
agricultural landownership is highly integrated into the sphere of
banking capital at a number of levels, yet such relations remain
obscured by analysis within the traditional frameworks of rural
sociology. Newby has drawn attention to this problem in a review
paper in which he argued the need for a distinctive 'sociology of
agriculture' and concluded that if this was to be met, 'the rural
sociologist must direct far more attention to landownership' (Newby
1982a).

Another factor in the lack of attention paid to landownership
in agricultural research has been the nature of the development of
landownership relations in British agriculture itself. The growth of
owner occupation as the majority tenure has ostensibly marginalised
rent as an economic category and with it any distinctive role for
landowners. It is significant that the recent revival of concern
amongst agricultural interests about the effects of landownership on
the industry has arisen largely as a result of the involvement of
the financial institutions in the agricultural land market. Their
involvement has revitalised the landlord-tenant system for the first
time since the war, and recreated an obvious and superficially
familiar expression of the rent relation.[2] They have also been
blamed for the massive increases in land prices experienced since
the 1970s which have been seen to present real problems for the
owner occupier (Norton-Taylor 1982, The Economist, 10 June 1978).
This interpretation derives from and perpetuates an ideological

perspective in which owner occupation is seen as an unproblematic, even ideal form of agricultural tenure.

In this paper I want to focus explicitly on the position of land and landownership relations in the development of capitalist agriculture. It is based on research which examined the significance of the growth of institutional landownership in British agriculture. In a limited but influential academic literature, much has been made of the theoretical importance of the growth of institutional landownership. Massey and Catalano's pioneering work, Capital and Land, presents institutional investment in agricultural land as a highly significant development both for the industry and in terms of Marxist rent theory. Although their book by no means focussed on the agricultural case, the significance they attached to institutional ownership of agricultural land has been eagerly taken up in less rigourous Marxist analyses of agriculture and by the political left.[3] Institutional ownership is interpreted not only as a 'logical' development of landownership relations under capitalism, but also as one which, over time, will come to dominate the agricultural sector, despite the contrary evidence of detailed empirical work on institutional purchases and holdings of agricultural land.[4]

I intend to reconsider the significance of institutional landownership in terms of a Marxist analysis of the historical development of agricultural production and landownership relations in Britain, or more precisely, in England over the last century.[5] In so doing, I want to address three related questions. First what is the material basis for the peculiar importance of landownership relations in the production and distribution of agricultural surplus value? Secondly how is this 'landownership relation' realised in both economic and social forms and how have these changed over time? And thirdly how do we understand the emergence and activities of institutional landownership within this framework? My purpose is not to present landownership relations as an independent determinant or 'motor' of agricultural change, but rather to define their role in the wider structure of agricultural productive relations and to identify the changing and diverse, social and economic forms in which they are manifest.

THE RELEVANCE OF MARX'S RENT THEORY

The property rights governing the ownership and use of land have a
profound effect upon the development and structure of agricultural
production. The nature of this relation is quite distinctive in
agriculture as compared with other industrial processes because of
the unique position of land in the agricultural production process.
Land forms a necessary <u>condition of production</u> in most industrial
processes, in the sense of a physical location or space. In
agriculture, however, land also acts as a central <u>means of</u>
<u>production,</u> that is, it enters into the production process, which in
turn becomes partially embodied in the land itself. Marx made this
critical distinction in his discussion of the origins of rent in
different industrial processes:

> In (mining) payment is made for land not because it is the
> <u>element in which production is to take place, as in</u>
> <u>agriculture</u>, not as one of the conditions of production, as in
> the case of...a building site, but because it is a reservoir
> containing the use values to (be exploited) (<u>Theories of</u>
> <u>Surplus Value</u> part 2:2450). (My emphasis).

The quality of land as a means of production, its fertility in
a broad sense, is highly variable across space. As a result the
division of land into discrete holdings creates <u>unique</u> parcels of
land in terms of its 'fertility'. Two important implications for
agricultural production derive from this essential characteristic.

First, individual capitals (farmers) can extract different
rates of profit from different land holdings independent of the
level of capital invested in them. Surplus profit is thereby
determined not by average prices of production but by those at the
margin (<u>i.e.</u> the holding with the lowest return for a given level of
investment)(Fine 1978, Ball 1979). It is this feature of
agricultural production which forms the main basis for the
extraction of ground rent from surplus profits under a system of
absolute property rights over the land.

The nature of land as a means of production in agriculture also
entails that it absorbs or embodies part of the capital invested in
the production process. Investments such as drainage, irrigation and
nutrient supplements become, to varying degrees, an integral element

of the land's fertility. Thus, much of agricultural capital can only be realised as surplus value by entering the land itself, which under a system of private landownership necessitates its entanglement in the rent relation.

Under a system of exclusive property rights, these characteristics of land as a means of production in agriculture confer a 'natural' monopoly power on the owner of the land. Owners have exclusive rights to the unique use values of their holdings and the power to appropriate a part of the surplus produced by any exploitation permitted on them. The dominant configuration of landownership relations observed by Marx constituted a distinct social class of landowners - Landed Property - which extracted ground rent in the form of a money payment termed rent, from a class of capitalist farmers, generally on annual tenancies.

As well as a striking analysis of nineteenth century British agriculture, Marx's theory of rent provides us with the conceptual tools with which to analyse the ways in which private property rights over land modify the formation and distribution of agricultural surplus value in the abstract, and in its contemporary social and economic forms. The importance of Marx's theory of rent is three-fold. First, it recognises rent as an economic expression of an essentially social relation derived from the institution of private property rights over land, rather than a neutral payment to an objective factor of production, land itself. Secondly, it shows how the rent relation affects not only the distribution of agricultural surplus value between classes, but actually modifies the productive relations of the industry and the way in which future surplus value is produced (Murray 1977). Thirdly, it establishes the essentially dynamic nature of the rent relation, recognising that particular social and economic forms of the relation are historically and geographically specific.

Many of the contemporary applications of this theory in the urban context have suffered from a preoccupation with the particular categories of rental expropriation devised by Marx, rather than with his method of analysis. Since these categories were devised to describe mechanisms of rental expropriation specific to nineteenth

century British agriculture, their simple transposition to
contemporary urban contexts has been fraught with problems. Even for
agriculture, both the internal productive relations of the industry
and its position in the wider national, political economy are vastly
different to those Marx observed. Seeking examples of modern
mechanisms for expropriating rents which mirror Marx's theoretical
categories of differential rents 1 and 2, absolute rent and monopoly
rent is, therefore, a barren route for research. A more fruitful
approach is to examine different ways in which the rent relation is
realised under a variety of forms and rates of accumulation in the
concrete social circumstances in which it is occurring (Ball et al
1984:4).

Marx's conception of the rent relation presents a considerably
more complex picture of the essential contradiction between private
property and capital than that of rent as a simple subtraction from
surplus profits, with solely negative effects on accumulation.
Rather, as Ball (1977) has emphasised, the rent relation
perpetuates the conditions which enable the continued existence of
surplus profits and in this sense, the basis of the existence of
rent itself. David Harvey (1982), in Limits to Capital, described
the relationship between landowners and capitalists as one of
'necessary compromise'. In acting as a barrier to the free flow of
capital onto the land, the rent relation rationalises the pattern of
production by concentrating investment and production into those
holdings and processes with the most profitable return. In practice
of course, this compromise is only rarely achieved. In a crude
sense, if too high a rent is extracted it erodes the capital
available to the farmer for investment and undermines profitability
and with it the basis of future rent. If too low a rent is
extracted, the rationalising effect of the rent relation is
weakened. In short, the role of the rent relation as a condition of
accumulation in agriculture is itself bound up in the central
contradiction between capital and private property rights.

In a very real sense then, the evolution of capitalist
agiculture has been critically influenced by the need to break the
dependence of production and accumulation on the land and to
circumvent the contradictions arising from the rent relation. Three

main strategies have been pursued to this effect.

1. The substitution or standardisation of land as a means of
production. Massive state expenditure on agricultural research and
direct subsidy to capital intensification has underpinned the
strategy of capital replacement for land and labour which has
characterised the development of post-war agriculture. In the case
of some sectors, factory farming and greenhouse production methods
have successfully replaced land as an essential means of production.
Whilst in others, particularly arable farming, effort has been
concentrated on standardising the 'fertility' or productive capacity
of the land by injecting synthetic supplements (e.g. fertilisers and
mineral supplements) or by modifying the soil requirements of crops
by genetic engineering.

2. The restructuring of landownership relations to avoid the
immediate barriers to investment presented by the prevailing form.
The growth of owner occupation as an alternative to the
landlord-tenant system appeared to provide a solution to the highly
visible and politicised conflict between the high rents charged by
landowners against the falling income of tenant farmers between the
1870s and the First World War. Likewise, more recently, the revival
of this distinction between landowner and capitalist farmer, with
the emergence of the institutional landowner, has been seen in some
quarters as a solution to the problems facing the owner occupied
farmer in the form of high land prices. These two developments in
the social structure of agricultural landownership in Britain are
amongst the most significant since the First World War. Yet,
critically, neither of them has resolved the contradiction of the
rent relation, but rather transformed the way in which it is
realised. This is not to suggest that there has been a linear,
historical development of these new forms of the landownership
relation. The complexity of tenure arrangements within farms as
business units has increased since the growth of both owner
occupation and institutional ownership. Mixed tenure arrangements
which have been widely adopted, combine various forms of the rent
relation in order to maximise access to land in a context of the
concentration and expansion of accumulation within fewer, larger
holdings.

3. The valorisation of agricultural products outside the farm-based stage of the production process, in the manufacture of agricultural inputs and the processing and packaging of agricultural products. The constraints placed on accumulation in agriculture by land as an awkward means of production, monopolised under private property rights, have discouraged the widespread direct subsumption of the farm-based stage of production by more corporate forms of capital. Instead large corporations of industrial capital have come to monopolise the expanded sectors of food processing and packaging and the production of inputs into agricultural production, particularly chemical and machinery manufacture. These stages of production have come to represent a far higher proportion of the value of agricultural products than farming itself.

THE COMMODITISATION OF LAND RIGHTS AND THE GROWTH OF OWNER OCCUPATION

Fundamental to the restructuring of landownership relations was the transformation of property rights over the latter half of the nineteenth century. Offer (1981), in his work on the development of British property relations, details the interventions of the state in this period, through which property rights were transformed, with some difficulty, into what MacMahon has termed 'quintessential objects of exchange' (1982:1). For the first time, land rights were freely and exclusively attainable on the open market as purchasable commodities.[6] It is worth emphasising that it is not land as a physical object, but titles to exclusive rights in land which are exchanged in the land market (Macpherson 1978). In this way land rights have become a form of 'fictitious capital'; that is, a paper title to the anticipated use values and revenue flows from the future exploitation of the land (Harvey 1982). As such land rights have acquired a price - representing at any one moment in time the capitalised value of anticipated future rents.

The commoditisation of property rights has transformed land into a more mobile and accessible form of capital, but one in which landownership is closely bound up with the contradictions inherent in the circulation of 'fictitious' forms of capital. These include stocks and shares, titles to property and land, and money itself.

They are by definition speculative forms of capital, whose price fluctuates not just in relation to the movement of the real agricultural use value of the land (in the case of land titles) but according to processes internal to the sphere of circulation: such as interest rates, the supply and demand for such titles as against alternative investments and the wider economic climate (Harvey 1982). Although this has made agricultural land a more mobile asset it has also separated agricultural land prices from the strictly agricultural use value of the land. This poses a number of constraints on agricultural accumulation for the owner occupier or any farmer wishing to buy land to increase his unit of production.

The social and economic relations of modern agricultural landownership have thus become thoroughly enmeshed in the sphere of finance or banking capital in which fictitious capitals circulate. As a result landowners no longer constitute a distinct social class in the Marxist sense, but rather a diversity of capitalists, including non-agricultural capitals, which have come to hold interests in agricultural land. Since the early twentieth century, the state has encouraged farmers (agricultural capitalists) to themselves invest in farmland as an alternative to the tenancy system (Sturmey 1955). Owner occupation has since grown to become the dominant tenure in British agriculture. Outwardly, owner occupation appears to combine the interests and roles of capitalist and landowner, thereby resolving the contradictions of the rent relation. In fact it has only been transformed through the mechanism of the land market. An appreciation of this is not new. Many of the Liberal land reformers and the tenant farmers being encouraged to buy their farms in the early twentieth century were apprehensive about owner occupation, because they foresaw large amounts of vital business capital being sunk into land purchase. The 1913 Rural Land Enquiry received plenty of submissions of evidence to this effect. With hindsight we can now see that the farmer who invests in land as an essential means of production for his business experiences the contradiction of the rent relation in a number of concrete ways, derived from the nature of land rights as a fictitious form of capital and from the very combination of business capital and landownership which defines owner occupation.

First of all, agricultural capital is forced to buy land rights
in a market where the price of land does not simply reflect its
agricultural use value. Moreover, owner occupiers cannot fail but to
contribute to the escalation in land prices. Whatever the validity
of Massey and Catalano's proposition that the purchase of land by
owner occupiers (as 'industrial landowners') is 'dominated by
considerations of the relevance of particular characteristics of
land in the process of production' (1978:64), they must still
operate in the same market as any other buyer, where the use value
and exchange value of land are inextricably linked. Their demand for
land places as much speculative pressure on land prices as does that
of any institution or 'investment' purchaser. Indeed, as
experienced, agricultural industrialists, owner occupiers may in
many ways be seen as market leaders in agricultural land. Their
willingness to purchase land reflects their confidence in the
industry's profitability and thereby provides a 'grassroots' guide
for institutional and other 'investment' purchasers as to the
security of land as an asset (Munton 1984:174).

The fact that agricultural production in the owner occupied
sector of the industry appears to be adversely affected by high land
prices is evidence of a facet of the contradiction between private
property rights and agricultural accumulation resulting from the
particular configuration of landownership relations found amongst
owner farmers. As Marx postulated,

> the price of land....a form and result of private ownership of
> land, itself appears as a barrier to production.
>
> (Marx - Das Kapital part 4:44).

As agricultural capitalists and landowners, owner occupiers are
active (and not always unwitting) participants in the speculative
rise in land prices, rather than the passive victims of outside
speculators or of a land market with a mind of its own.

A second aspect to this contradiction concerns the capital
liquidity crises often experienced by owner farmers. Surplus profits
required for investment and working capital in the production
process must, under owner occupation, extend to purchasing the
freehold rights to the land. Sinking this amount of capital in land

rights can seriously detract from the liquid capital available for productive investment and business development. Whilst wealthy on paper, the owner occupier often faces a continuous cash-flow problem. This problem is exacerbated by the high price of land mentioned above and by the highly individualised nature of capital ownership peculiar to agricultural businesses, which provides little recourse to outside sources of capital, bar credit.

This brings us to the third element of contradiction of the rent relation realised under owner occupation - the use of land as the basis of agricultural credit relations. The commoditisation of land rights together with the problems faced by the farmer in the purchase of land has increased the role of the short term lending institutions of banking capital in agricultural landownership and productive relations. Two main mechanisms can be identified. First, the use of land rights as the chief form of collateral against a loan of liquid capital for investment (or personal consumption). Under this arrangement, the owner occupier's land rights are held by the lending institution as security against a loan. Secondly, the purchase of land rights through a mortgage arrangement. In this relation, the nominal owner-farmer surrenders the title of the freehold land rights to a lending institution in return for a money loan on the condition that the title shall revert on the payment of the loan on fixed terms. In either instance, nominal owner-occupation disguises a credit relation equivalent to a form of rent. Figure 1 illustrates the growth in the proportion of agricultural surplus taken up by interest payments from farming capital. As Shalit and Schmitz (1982) have argued, the debt carrying capacity of farmland - determined largely by the willingness of the short term lending institutions of banking capital to accept land as collateral for credit - has itself become an increasingly important influence on the market value of agricultural land.

These economic realisations of the rent relation within 'owner occupation' present real barriers to capital investment in agriculture and materially affect the process of accumulation in the industry. Their effect has been considerably exacerbated by the form and scale of post-war support for agricultural prices and subsidy to capital investment. By underwriting agricultural profitability, the

FIGURE 1

INTEREST BURDEN ON UK FARMING 1970-1983

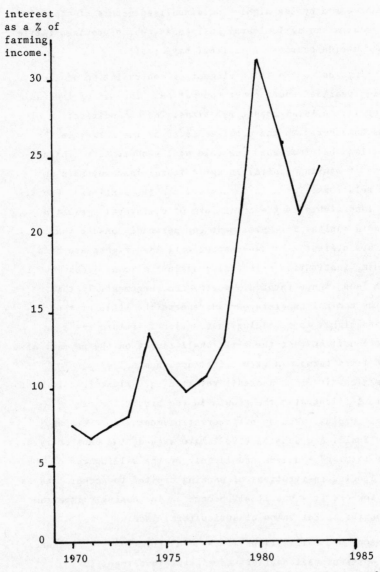

interest
as a % of
farming
income.

1 Interest refers to interest on commercial borrowings excluding
loans for land purchase.
2 The index is constructed on the basis of constant (1975) prices
and current prices.

From Burrel, A., Hill, B., Medland, J. 1985. p66.

state's agricultural policies have underpinned 'confidence' in the speculative or investment market in agricultural land rights, which in turn has fed into high land prices. Moreover, the absorption of much of the subsidised capital inputs into the land has fed directly into increased rents and higher land values. Traill's model (1980, 1982) of the transformation of state agricultural subsidies and supports into rent and land prices is of interest here. He calculates that a notional 1 per cent increase in support prices produces an initial increase of 8 per cent in profits on farm incomes, and a 10 per cent rise in rents and land prices after 1-2 years (Traill 1980:32).

These constraints presented by landownership relations within owner occupation underline the importance in the survival of farm businesses of non-market methods of land transfer, such as inheritance and succession (Delphy 1984), and the development of more corporate forms of both family and non-family farm businesses as a means of transfusing new sources of capital (Marsden 1984). The transformation of land rights into financial assets and the development of the land market as a specialised investment sector has also created the conditions within which the institution of finance capital might increase their involvement in agricultural land, not just indirectly through the credit relations mentioned above, but directly as landowners.

THE SIGNIFICANCE OF INSTITUTIONAL LANDOWNERSHIP

The sector of finance capital which has invested directly in agricultural land rights over the last 15 years or so, is the long-term investment institutions, including: pension funds, insurance companies and property unit trusts.[7] The rationale behind institutional landownership lies in the particular pattern of development of finance capital in Britain in the post-war period and cannot be interpreted in a simple, functionalist way as a 'necessary' development of agricultural landownership relations (Whatmore 1983 chapter 6). Nevertheless they have come to represent an important means by which agricultural capital can be released from land held under owner occupation.

The growth of institutional investment in agricultural land in the late 1960s can be seen as the result of a combination of particular historic conditions. Nationally, the economy was characterised by very high rates of inflation, government restrictions on overseas investment and the poor performance of traditional long-term investments such as government stocks. The investment institutions themselves were experiencing an unprecedented growth in funds (much of it with a long-term liability profile) derived from the savings boom of this period. Together with the restricted options on traditional investments these conditions encouraged the institutions to look for new, long-term sectors of investment, in particular, property of all types. Finally, agricultural land rights appeared as a particularly secure, long-term, inflation-proof investment because of the long standing state support for agricultural prices and profits.

Two features of institutional investment in agricultural land stand out. First is the highly selective pattern of investment, confined to and increasingly defining a 'prime' agricultural land market. This is characterised by holdings on the top grades of land; farm units over 350 acres; and predominantly arable farms. The units are also predominantly tenanted, although sale and leaseback arrangements and, to a lesser extent, in-hand farming have increased the institutions' involvement in the vacant possession land market. In other words, the 'prime' market refers to qualities both of the land as a means of production and the farm business as a profit generating concern. The second feature is the high management profile of institutions resulting from their growing interest in the income performance as well as the capital appreciation of their agricultural land investments. Thus, as landlords, they have typically adopted rent maximising strategies involving both financially stringent rent agreements and active investment in fixed farm capital - realised in returns as increased rent or as credit repayments from the tenant farmer. In some cases they have entered more directly into the agricultural business itself thereby diversifying their investment income from strictly rent, to profit and interest on credit. This has been achieved either through partnership arrangements, commonly with an established farmer, or through taking land in-hand and managing it through a subsidiary farming company.

In some ways the development pattern of institutional involvement in agricultural land mirrors the development of their role in the urban property market; acting first as a source of long term loan capital to developers, then establishing partnership arrangements with development companies and finally absorbing the development business into their own organisation as a subsidiary concern (Barras and Catalano 1975). There are, however, two significant differences. First, instead of large corporate, public companies which characterise the development industry, agricultural businesses are predominantly family based, even if they have adopted more 'corporate' forms such as limited companies, partnerships, trusts or syndicates. Secondly, the institutions' holdings of agricultural land are much less important than their non-agricultural property assets, in terms of both the significance of agricultural holdings against their total assets and the extent of their holdings as a proportion of the total market.

Nevertheless, the significance of institutional landownership for the development of the agricultural industry has, I would argue, been widely underestimated, particularly by the Northfield Committee[8] and opinion informed by its report, precisely because its impact has been measured purely on the basis of the percentage of total agricultural land held by the institutions (which is less than 2 per cent). Other features of institutional ownership would suggest a larger significance.

First, their holdings are concentrated on prime agricultural units, which are much more highly capitalised and larger than the average, and therefore account for a far higher proportion of agricultural output and capital assets than their areal extent would suggest. Moreover, their sectoral and locational position is considerably more dominant than in agriculture overall, being concentrated on arable holdings in Eastern England and South East Scotland (Savills/RTP 1982, 1983, 1984). They have also built up this position over just a short period of a few years.

Secondly, the institutions acquire and manage agricultural land rights on the same basis as any other financial asset. In this sense, they represent the most 'highly developed' form of capitalist landownership. Although some more traditional institutions may invest on a similar basis, their agricultural landholdings are a far more dominant element of their total investment portfolios (Munton 1985). Financial institutions, therefore, represent the most highly volatile influence in the agricultural land market. Their strictly financial interest in agricultural land and its marginal position within their total asset structure makes institutional investment extremely sensitive to any fall in income or capital growth. Already, in the wake of a deceleration of rental growth and a lower rate of inflation, the institutions have become more cautious and less active in the acquisition of agricultural land. Crucial to their future investment in agricultural land is the stability of state support for agriculture, as this is fundamental to the performance of agricultural land as an investment asset. With both the Government and the EEC reconsidering their policies of unconditional agricultural support and extensive subsidy arrangements, the institutions have curtailed the purchase of new holdings and rationalised their existing holdings. Over 25 per cent of total institutional acreage, some 10,000 acres, mainly in Eastern England, was reported to be on the market during the summer of 1985.[9]

Thirdly, the institutions have established a sale and leaseback market in agricultural land, whereby an owner occupier sells land to an institution in return for a lease agreement to farm the land, often under some partnership arrangement. More than 60 per cent of institutional properties have been purchased in this way, representing 45 per cent of their total acreage (Savills/RTP 1983, 1984). This represents a new and important means of releasing agricultural capital from the constraints of landownership relations under owner occupation. Fourthly and finally, the institutions represent an important new source of capital in the industry both through direct investment as partners and through subsidiary farm companies, and as a source of loan capital to tenant farmers.

The significance of the institutions' investment in

agricultural land is not as an isolated or independent force in the
agricultural economy, with discrete, quantifiable effects. Nor, I
would argue, can they be said to represent a distinctive new 'class'
of landowner either in the sense popularised by the press, of a new
'landed aristocracy', or in the stricter sense intended by Massey
and Catalano's definition of 'financial landownership'. Their
distinction between 'traditional', 'industrial' and 'financial'
landowners - represented in agriculture by the crown and the church,
owner occupiers,and financial institutions respectively - is
conceptually 'chaotic', to use Sayer's term (1979). It presupposes a
rigid separation between the speculative and use value elements of
agricultural land rights. Given the nature of land rights as a form
of fictitious capital, such a separation does not exist at a
theoretical level, nor can it be realised in this way as concrete,
social forms of landownership. The owner occupier cannot escape and
indeed, contributes to the speculative element of agricultural land
prices, and on paper at least is made wealthy by it. Similarly, the
institutions' interest is ultimately tied to the future agricultural
use value of the land, although it is purchased solely as an
investment.

CONCLUSION

Within the framework set out in this paper, the significance of
institutional landownership for agriculture can be seen to be
two-fold: as a step towards integration of agricultural
landownership into the sphere of banking capital; and in the
distinctive modification of owner occupied landownership relations,
through sale and leaseback arrangements. However, institutional
ownership is only one expression of the contradictions associated
with land rights as a fictitious form of capital. There are other,
more important, social forms, such as owner occupation and
land-based credit relations. It is important therefore to keep in
perspective any specific form whereby the central contradiction is
realised between private property and the capital accumulation
process in agriculture.

Thus, while focussing on the neglected issue of agricultural
landownership, this chapter has also emphasised its position within

the evolution of the wider productive relations of the industry, rather than isolating or reifying it. We certainly cannot properly understand the social organisation of agriculture, its characteristic business structure, or the importance of the family in modern agriculture, without comprehending how landownership mediates the use of land in agricultural production, and effects the production and distribution of surplus value in the process. Equally, however, in following Newby's prescription for a 'distinctive sociology of agriculture' we must be careful not to reduce the development and structure of agricultural relations to a simple function of the pattern of landownership.

FOOTNOTES

1 See Marx Theories of surplus value part 2 and Das Kapital vol.3, part 6; also Ricardo The principles of political economy and taxation, chapter 2.

2 The percentage of tenanted land rose for the first time since the First World War in the early 1970s when the institutions were actively purchasing tenanted estates.

3 Much of the criticism of institutional landownership by the political left is also populist rather than analytical. See for example papers given at the Socialist Land Conference held at the GLC in May 1985.

4 Detailed information is provided by the Agricultural Performance Analyses issued by the land agents, Savills, in conjunction with Roger Tym and Partners; Jones Lang Wooton 'The agricultural land market'1983; and Munton 1985.

5 The historical development and contemporary pattern of agricultural landownership in Scotland and to a lesser extent Wales are quite distinctive from that in England. It is worth noting that South East Scotland, the heart of that country's agricultural sector, has become a major location for institutional investment in the U.K.

6 See F.M.L. Thompson's evidence to the Northfield Enquiry for a concise history of the development of the English land market.

7 For a discussion of the development and wider activities of the financial institutions see the Wilson Committee Report, 1980.

8 An official committee of inquiry, chaired by Lord Northfield was set up in 1977 amidst political concern over rising farmland prices and the growing involvement of financial institutions in the rural land market. The background to the establishment of the committee and an analysis of its work is given in Munton 1984.

9 Information gained from interview with Savills the land agents in July 1985.

REFERENCES

Ball, M. 1977. Differential rent and the role of landed property. Int. J. Urban Reg. Res. 1, 380-403.
Ball, M. 1979. On Marx's theorty of agricultural rent: a reply to Ben Fine. Econ. & Soc. 9, 304-326.
Ball, M., Bentivegna, V., Edwards, M. and Folin, M. 1984. Land rent, housing and urban planning. London: Croom Helm.
Barras, R. and Catalano, A. 1975. Investment in land and the financial structure of property companies. London: Centre for Environmental Studies.

Burrel, A., Hill, B. and Meland, J. 1985. A statistical handbook of UK agriculture. London: Macmillan.

Delphy, C. 1984. Close to home. London: Hutchinson.

Fine, B. 1978. On Marx's theory of agricultural rent. Econ. & Soc. 8, 241-279.

Harvey, D. 1982. Limits to capital. Oxford: Basil Blackwell.

Jones Long Wooton. 1983. The agricultural land market. London.

Macpherson, C. B. 1978. Property: mainstream and critical positions. Toronto: University of Toronto Press.

McMahon, M. 1982. A general theory of rent. Mimeo. London: Bartlett School, University College.

Marsden, T. 1984. Landownership and farm organisation in capitalist agriculture. In Locality and rurality, T. Bradley and P. Lowe (eds), 129-145. Norwich: Geo Books.

Marx, K. 1964. Theories of surplus value part 2. London: Lawrence and Wishart.

Massey, D. and Catalano, A. 1978. Capital and land. London: Edward Arnold.

Munton, R. 1977. Financial institutions and their ownership of agricultural land. Area 9, 29-37.

Munton, R. 1984. The politics of rural landownership: institutional investors and the Northfield Inquiry. In Locality and rurality, T. Bradley and P. Lowe (eds), 167-178. Norwich: Geo Books.

Munton, R. 1985. Investment in British agriculture by the financial institutions. Sociologia Ruralis 25, 155-173.

Murray, R. 1977. Value and the theory of rent part 1. Capital and Class 4, 100-122.

Murray, R. 1978. Value and the theory of rent, part 2. Capital and Class 3, 11-33.

Newby, H. 1982a. European social theory and the agrarian question: towards a sociology of agriculture. Paper presented to Rural Economy and Society Study Group, Oxford.

Newby, H. 1982b. Rural sociology and its relevance to agricultural economics: a review. J. Agric. Econ. 33, 125-165.

Northfield Committee. 1979. Inquiry into the acquisition and occupancy of agricultural land. Cmnd 7599. London: HMSO.

Norton-Taylor, R. 1982. Whose land is it anyway? Wellingborough: Turnstone Press.

Offer, A. 1981. Property and politics 1870-1914. Cambridge: Cambridge University Press.

RTP/Savills. 1982. Agricultural performance analysis. London.

Shalit, H. and Schmitz, A. 1982. Farmland accumulation and prices. Am. J. Agric. Econ. 64, 710-719.

Sturmey, J. G. 1955. Owner farming in England and Wales 1900-1950. Manchester School 23, 246-268.

Traill, W. B. 1980. Land values and rents. Bulletin 175. University of Manchester: Department of Agricultural Economics.

Traill, W. B. 1982. Taxes, investment incentives and the cost of agricultural inputs. J. Agric. Econ. 33, 1-12.

Whatmore, S. J. 1983. Financial institutions and the ownership of agricultural land. M. Phil. Thesis. University of London.

Wilson Committee. 1980. Review of the functioning of financial institutions. London: HMSO.

Winter, M. 1984. Agrarian class structure and family farming. In
    Locality and rurality, T. Bradley and P. Lowe (eds), 115-128.
    Norwich: Geo Books.

8 Property–state relations in the 1980s: an examination of landlord–tenant legislation in British agriculture

TERRY MARSDEN

INTRODUCTION

This paper examines the changing relations between landowners and their tenants in Britain during the post-war period. An understanding of the significance of landownership and landlord-tenant relations increasingly necessitates a focus upon the interface between a capitalist land market and the spasmodic, and often contradictory intervention of the state. The discussion here is divided into three parts. First, an examination of property-state relations is made with regard to British agricultural land. Secondly, a largely empirical section details the changes that have occurred in agricultural landownership during the twentieth century, paying particular attention to the changing nature of the landlord-tenant system of occupation. Finally, state policy affecting landlords and tenants is assessed with specific reference to the passage of the 1984 Agricultural Holdings Act. The advent of this piece of legislation represents a particular episode in property-state relations. The primacy given by the state to maintain the flexibility and autonomy of landowners in the 1980s can be contrasted with earlier attempts to constrain their control of land through increasing tenant security.

PROPERTY-STATE RELATIONS AND THE CONTROL OF AGRICULTURAL LAND

Societies require a degree of consensus concerning property rights. Laski (1934), for example, regarded it as essential for there to be a sound justification of property for society as a whole, though he rejected efforts to do this in psychological, scientific and ethical terms. His own justification combined meeting the minimum needs of individuals with criteria for their personal efforts and social worth. Macpherson (1973) argues, however, that progress towards a truly democratic society - one which enables individuals 'equally to use and develop their human capabilities' - is thwarted by the traditional institution of property. Its exclusivity, in capitalist societies at least, is underpinned by the utilitarian principles of the right to possess, use and dispose of the fruits of ones own exertions (Locke 1960). The exclusive and often monopolistic character of property hinders the egalitarian goal of maximising individuals' development powers by maintaining 'extractive power' i.e. the ability to use the capacities of other men.

Under the conditions of developed capitalism, as in contemporary Britain, this state of affairs has been substantially modified by the state. Since the turn of the century, for instance, exclusive property rights have been seen as a right to revenue by governments. In some cases the nationalisation of royalties has removed the rights of landowners to the rental income from the minerals (e.g. coal) held beneath the land they own. In agriculture, of course, no such removal of individual property rights has occurred despite a series of attempts to tax the revenue of private owners, particularly on the transfer of assets from one generation to another. The contradiction between democratic rights (and the equalisation of property to some degree) and the utilitarian notions of private property as a reward for effort is long established in British society. The political expression of democratic principles (outlined by such theorists as Laski and Macpherson) confronts capitalistic processes which necessitate both the commodification of land rights and their concentration in the hands of a minority of individuals and institutions (see Whatmore Chapter 7).

In many ways private property rights in agricultural land stem from the historical conditions in which capitalism became the dominant mode of production. Private landownership was the basis upon which the conditions were established for the rise of capitalism. Landed property was not, as under feudalism, to be the

basis of class antagonism, but its function as a necessary condition
of production, combined with unlimited rights, conferred upon its
owners considerable political and economic power (see Massey and
Catalano 1978). Modern ideological justifications for private
property have their roots in this transition (see Veblen 1923).

Since the nineteenth century, private property rights have
become a continuous source of conflict in British society as the
class interests excluded from that ownership have challenged the
absolute rights to property and the wealth that accrues from them.
Certain forms of ownership have been particularly contentious, not
least the extensive ownership of land and the rights it confers to
appropriate rents, as well as the use of succession to perpetuate
ownership. Indeed, the inheritance of wealth in the form of tangible
commodities, particularly land has created considerably more
political tension than, for instance, the concentration of finance
capital. The structural contradiction which thus develops between
the tendency on the one hand towards the monopolisation of property
rights and the problem of legitimacy this creates for societies with
democratic aspirations, can only be mediated by the increasing
intervention of the state. The growth of capitalist relations of
landownership provides the state with a major project, that of
legitimising the concentration of property rights amongst a
minority, or at least mediating between economically powerful
landowning interests and the rest of the class structure who are
challenging such monopolistic tendencies.

The state, coherently or otherwise, must repeatedly produce
strategies for action which will justify the maintenance of private
property rights. Over time we can record that the state has acted in
a variety of ways to modify, concentrate or even destroy private
property rights, from the threat of land nationalisation to the real
taxation of unearned income. The significance of state action is
not, as in many other areas of involvement, unidirectional. Rather,
it has attempted, usually with a short term perspective to mediate
tensions stemming from the contradiction between private property
and democratisation when and where they have arisen, protecting the
autonomy of landowners on the one hand, and then embarking upon
sporadic political attacks on the landowning faction on the other.

Bornstein et al (1984) have aptly identified the contradictory roles
of the state:

> On the one hand the state must sustain the process of
> accumulation and the private appropriation of resources, on the
> other hand, it must pursue belief in itself as an impartial
> arbiter of class interests thereby legitimating its power.
>
> (Bornstein et al 1984:16).

The primacy of private ownership and control of property has
thus been maintained. Through increasing emphasis on the efficiency,
stability and expansion of agricultural production within the
established landownership structure, the state has reinforced the
justification of private property rights in the post-war period.
Rights over agricultural land are today justified not only in social
and political terms but by economic criteria associated with the
maintenance of an 'efficient' food producing system which is, of
course, as much a product of state action as it is of the workings
of the market. Private property rights are increasingly justified
according to their 'accumulative potential' (see Newby et al 1978).

The state has promoted these conditions for the full
exploitation of property rights through agricultural production
mainly by guaranteeing prices of goods produced, which has also
stimulated, artificially, an increase in the price of land itself
(see Trail 1981). On the other hand, the state has embodied
political attacks upon landowners. As we shall see below, the
attempts to tax the transfer of assets through Wealth and Capital
Transfer Tax, and particularly to confer upon tenants increased
security of tenure and allow family continuity on tenanted farms,
have been significant in reducing the owners' control of land. Such
threats to private landownership have elicited a high level of
political mobilisation on the part of farmers and landowners through
their respective interest groups - the Country Landowners'
Association (CLA) and the National Farmers' Union (NFU). These have
attempted to protect traditional property rights when and where they
have been threatened; their opposition to post-war planning controls
being perhaps the most prominent example (Newby et al 1978, Lowe
et al 1986).

The succeeding discussion will demonstrate that while the need

to protect landowners' freedom to act on land they control can
obviously be viewed as one reason for this mobilisation, a more
fundamental need is to maintain the accumulative potential of their
property and the flexibility to modify productive systems to ensure
efficient returns on capital invested. After outlining some of the
most significant changes occurring in the landownership structure
since the turn of the century, discussion focuses upon the changing
role of state action in modifying landlord-tenant relations.

CHANGES IN THE STRUCTURE OF AGRICULTURAL LANDOWNERSHIP AND THE SURVIVAL OF THE LANDLORD-TENANT SYSTEM

Of agricultural land in Great Britain, 8.5 per cent is owned by
public institutions, 1.2 per cent by financial institutions and the
rest (90.3 per cent) by private individuals, companies and trusts
(see Northfield 1979, Jones Lang Wootton 1983). There are serious
problems in disaggregating the largest category due to the lack of
official data on landownership in the United Kingdom. Private
individuals may own land through a variety of different
partnerships, companies and trusts and they may also own or let land
solely or jointly with someone else. The formation of trusts is
usual for large areas of let land while for smaller areas of land
sole or joint ownership predominates. It is important to ascertain
the proportions of the privately owned land which is controlled by
owner-occupiers and by private landlords as there are considerable
differences between the groups in their degree of control and ways
of organising production. The number of owner-occupied farms has
increased throughout the century such that by 1983 entirely owner-
occupied farms accounted for 57 per cent of all holdings and 41 per
cent of all agricultural land. If one includes those holdings that
are mainly owner-occupied the figures rise to 62 per cent of all
holdings and 57 per cent of the total area of farm land (Northfield
1979). The remaining 30 per cent of privately owned land is in the
hands of the private estate landlords who may let all the land they
own or farm some of it themselves.

As a result of state action and market forces the pattern of
landownership has changed considerably since the beginning of the
century. Owner-occupation is estimated to have been as low as 10 per

cent of all agricultural land throughout the period 1850-1914
(Thompson 1979). A major structural shift occurred during the 1920s
when, as Thompson (1979:29) records: 'The land market exploded with
an unprecedented frenzy of activity...in the space of two years
(1921-22) the ownership of more than a quarter of agricultural land
of the country changed hands'.

Legislation giving the security of tenure to tenants
(particularly the Agricultural Holdings Act of 1908), the decreasing
levels of return upon fixed capital and the gradual increase in the
price of land with vacant possession all gave strong incentives to
private landlords to sell off their estates either in part or as a
whole. Sturmey (1955) argues that state action in the early 1900s
unfavourable to the landowning classes was particularly significant
in redistributing the ownership of agricultural land among a larger
group of owner-occupiers:

> The combination of rising land prices and low rents which led
> landowners to sell their properties in the years before 1909
> continued after 1909, supplemented by a campaign which moved
> from a mere vilification of landlords as a class to threats of
> nationalisation of land by the imposition of heavy taxes
> thereon.
>
> (Sturmey 1955:252).

From 1919 onwards, new methods of calculating land values were
imposing heavier death duty payments upon landowners. This further
induced them to sell their estates often to existing tenants. For
Patrick Abercrombie, the architect-planner who founded the Council
for the Preservation of Rural England, these changes in the
landownership structure were responsible for the rapid deterioration
of the traditional rural landscape and the encroachment of urban
sprawl:

> even where the estates remained in tact the rapport that had
> existed between local inhabitants and the previous owners had
> been destroyed. The traditional feeling of being jointly
> responsible for preserving local amenities had gone.
>
> (Abercrombie 1930).

In addition, many of the heavily mortgaged, former tenant farmers
were hard hit by the dramatic fall in agricultural prices in the
1920s and 1930s and were forced to sell off land.

A second surge of activity occurred during the post-war period, 1950-60, which acted to further increase the proportions of owner-occupied land from about 35 per cent to over 50 per cent. Increasing taxation and the rising costs of maintenance were encouraging the further break up of estates and the reduction of the 'let' sector of the land market. Rent regulations under the 1947 Agriculture Act and the 1948 Agriculture (Scotland) Act, while increasing the security of tenure for tenants (see below) acted further to induce the landlords to sell their farmland or to take land in hand. A third period of change has occurred since 1970, typified by general price inflation, large fluctuations in the price of agricultural land, significant rises in farm rents and radical changes in tax regimes. The growth of owner-occupation has continued but with the specific increase in financial institutional investment.

Although the financial institutions have been predominantly concerned with acquiring land which is then let to tenants, their growing interest in land with vacant possession and direct farming was seen as a worrying trend by the Northfield Committee (1979). By 1978, 19 per cent of the institutions' land was farmed directly (i.e. in hand or in partnership). The Committee, appointed by the Labour Government to investigate the consequences of institutional ownership, was chaired by the Labour peer, Lord Northfield. It estimated that, by the year 2020, over 15 per cent of the total agricultural area could be in the control of public and private institutions. Despite these trends the period since 1970 has been characterised by a slowing down in the rate at which land has changed hands. By the late 1970s only 1-1½ per cent of all agricultural land (180,000-270,000 hectares) came onto the market annually, compared with 3 per cent in the early 1950s.

Despite the incursion of financial institutions into the land market throughout the 1970s (Munton 1977, 1985, Roger Tym 1983) and the considerable attention this has attracted regarding their possible marginal interests in the continuity of farming and landscapes, a major feature of this contemporary period has been the continuance of the traditional private estate landlord. By 1978, 43 per cent of all agricultural land and 37 per cent of all holdings

were rented although this overestimates the true extent of
landlordism because, in certain cases for tax advantages, a farmer
is the nominal tenant of a farm company of which he is the
beneficial owner. Although the public institutions and the majority
of financial institutions do let out their land, private landlords
still remain the most significant guardians of the tenant farming
system. Their role in maintaining the 'let' sector was endorsed by
the Northfield Committee in 1979. They argued that the role of the
financial institutions in agriculture ought to be as long-term
investors in let land and not as farmers and that there was no
reason to 'hasten the departure of the good private landlord'.
Echoing Abercrombie some thirty years previously, their continuance
was seen as crucial to the environmental and social stability of
rural society. In terms of the Northfield enquiry the 'accumulative
potential' of private landlords and their flexibility to modify
production systems should be encouraged because:

> The good private landlord has brought great benefit to farming,
> the countryside and the quality of rural life. Through a sense
> of 'stewardship' he has worked actively to create and preserve
> beauty of the landscape and has exercised social responsibility
> for local communities.
>
> (Northfield Committee 1979:210).

Rejecting, on agricultural and efficiency grounds the
nationalisation of agricultural land, major support and
encouragement was given to the private landlord-tenant system for
the purposes of maintaining movement in and out of agriculture by
tenants and in generally developing a more competitive, diversified
land market. Of particular concern was the decline in the number
falling vacant for letting, to less than 2 per cent of all tenancies
each year.

In the late 1970s and early 1980s, the major areas of debate
have focused upon the problems of encouraging a vibrant
landlord-tenant system rather than on the curtailment of property
rights of landowners themselves. As in other areas of state
involvement the focus of concern has shifted from that of positive
intervention towards ways of encouraging landlords to maintain their
properties and their custodial role in rural society. In this way,
the existing structure of landownership is endorsed. An examination

of recent legislation concerning the security of tenure and
succession rights of tenants demonstrates changes in state policy,
which allow the maintenance of existing landownership structures.

LANDLORD-TENANT RELATIONS AND THE QUESTION OF CONTROL

Apart from going out of agricultural use, land has been lost from
the tenanted sector in two major ways. Landlords continued to sell
off their land to existing tenants throughout the 1970s and early
1980s at the rate of 0.1 to 0.2 per cent of all tenanted land per
year. Also, and more significantly, landlords have progressively
farmed more of their estates themselves. A sample study conducted by
the Central Association of Agricultural Valuers (1978) for instance,
discovered that by far the largest proportion of land (75 per cent)
which fell vacant was taken under the control of existing landlords,
farmed in partnership, or eventually sold. While there is a tendency
to associate the decline of the tenanted sector with the decline of
the private landlord per se, the process of taking land 'in hand' or
developing trusts and partnership agreements amongst a smaller group
of tenants, suggests that the private landlords' significance in the
land market and in their relations to tenants has become much more
diversified.

Three major sets of factors can be established in explaining
these changing circumstances, and particularly the decline of the
traditional landlord-tenant system. First, increasing farm size, and
the tendency to amalgamate small and medium sized farms in the
interests of efficient estate management has, by default, reduced
the absolute number of tenanted farms as well as the proportion of
small farms to serve as the first rung on the agricultural ladder.
The Smallfarmers' Association, and the more recently formed Tenant
Farmers' Association, (established in 1981) have been pressurising
the NFU and the CLA to support the case for an expansion and
revitalisation of the tenanted sector, to facilitate entry into
farming.

Secondly, increasing fiscal pressures (particularly Capital
Transfer Tax) and investment income surcharge on rents, and the
threats of a Wealth Tax, all encouraged landlords to abandon their

traditional role of letting land. Thirdly, and of particular
significance here, the letting of land has been discouraged by
certain statutory restraints which, it is claimed, have reduced
landlords' freedom of management and, in particular, the freedom to
relet land when and to whom they choose. The landowners, through
their interest groups, have been at pains to demonstrate that any
types of restrictions imposed upon them are likely to reduce the
number of tenancies even further as well as disrupt relations with
their existing tenants. Their efforts to protect their accumulative
potential and control over the land have been justified by reference
to the need to maintain a buoyant tenanted system of occupation.

The post-war Labour administration placed much more emphasis
upon increasing the security of tenure of tenants, however. The
Agricultural Holdings Act (1948) ensured the security of tenants
during their lifetime. Where a landlord attempted to serve a notice
to quit to one of his tenants, the action needed the consent of an
Agricultural Land Tribunal. The tenant was responsible for
initiating this process by serving a counter notice. Any notice to
quit would be considered legitimate if one of a series of criteria
were substantiated; for instance, if it was judged that the tenant
had been practising poor husbandry in comparison with the landlord's
system of farming, or where the farm tenant was not fulfilling his
'obligation' to farm properly. If one landlord could prove this was
the case, the Agricultural Land Tribunal would then issue a
certificate of bad husbandry. Also, tenants could be removed to
facilitate 'sound estate management' which often meant allowing for
some form of amalgamation.

Tenants had little power to challenge the landlords'
initiatives in serving the notice to quit, but after the
Miscellaneous Provisions Act of 1976, the tenant was allowed to
demand that the Agricultural Land Tribunal consider whether the
landlords' claims were 'fair and reasonable'. A tenant who was
insolvent or who refused to pay rent could not mount such a
challenge. Whereas the 1948 Act in laying the foundation for tenant
security restricted landowners' powers to remove tenants and obtain
vacant possession, it was the 1976 legislation which was to be
perceived by landlords as an unwarranted intrusion into the workings

of the land market. It gave primacy to the continuity of family
farming rather than the maintenance of landowners'control over land
or the needs to revitalise the 'farming ladder'. Until 1976, the
death of a tenant provided the landlord with an opportunity to serve
an incontestable notice to quit within three months. The 1976 Act,
for the first time, enabled certain family members of the deceased
tenant to claim the tenancy. Succession would also be allowed to the
prospective third generation, before the landlord could terminate
the agreement. As a consequence, security of tenure was potentially
extended from one to three generations. Nevertheless, there were
considerable safeguards for the landlord built into this procedure.
The potential successor needed to apply for the privilege and
needed to be 'eligible'; he had to be related to the deceased tenant
and his principal livelihood must have been from the farm itself. He
must not have occupied - either as an owner or as a tenant - another
'commercial unit' (i.e. a farm capable of supporting the full
employment of two people). In addition to this the successor was
also required to prove his 'suitability' through age, health,
training and experience, and financial backing. Where several
potential tenants had presented themselves, the Agricultural Land
Tribunal was empowered to make the final choice.

While the 1948 and more particularly the 1976 legislation were
viewed as essentially anti-landlord measures introduced by Labour
Governments, there were rational agricultural reasons for increasing
security of tenure. It would encourage tenants to invest in
improvements which hitherto had usually been provided by the
landlord, and would also preserve the family farming character of
the businesses. Many tenancies would be preserved which otherwise
would have been taken 'in hand' by the landlord. The NFU was broadly
supportive of the Acts' intentions, while its sister organisation,
the Scottish Farmers' Union, and the separate Farmers' Union of
Wales strongly favoured the measure, reflecting the continued
prevalence of a degree of antagonism between a small tenantry and
large estate owners in the Scottish and Welsh uplands.

Nevertheless, the Act came under considerable attack from
landowners themselves and their representative organisation, the
CLA. In a context of rising land values and increasing demand for

farms landlords were encouraged to seek as much control as possible over their land and to protect their option on vacant possession. The landlords argued forcibly that the encroachment of the 1976 Act on their rights would further reduce the fluidity of the let sector, reduce the potential value of land by not allowing the vacant possession premium to be obtained, and act in general to discourage the landlord to re-invest in the maintenance of farm properties and, by implication, the rural environment over the longer term. Reducing their autonomy and flexibility to modify systems of production would further encourage 'in hand' farming by landlords and the proliferation of fixed yearly licences and other forms of short term lease. The NFU, which originally supported the legislation, soon joined with other interest groups in criticising its possible consequences. This change of policy was partly influenced by the continual fall in the number of tenancies and the lack of clear evidence concerning the possible advantages originally envisaged in the construction of the Act. Moreover, the gradual convergence of the NFU and the CLA's membership due partly to changes in farm structure (especially in-hand farming and increasing owner occupation) eroded support within the NFU for a measure that could be portrayed as inimical to private landownership. After 1979 and the change from a Labour to a Conservative Government, this expressed itself in a shift of emphasis away from the ideology of family continuity (embodied within the 1976 Act) and towards the need to ensure increases in the number of tenancies and the revitalisation of the farming ladder. This shift coincided with mounting criticism of overproduction in agriculture and the extent of public financial support for farmers. It became increasingly necessary for the farming lobby to dispel the ascendant image of agriculture as a closed and privileged profession.

Despite the original justification for increasing the security of tenure being to discourage tenants from 'taking the last half penny of profit out of the land during a period to which they have a contractual right to remain in occupation' (Verrall v. Farnes 1966), the major controversy focused upon the pro-tenant/anti-landlord character of the legislation. Along with the introduction of Capital Transfer Tax, landowners and most of the agricultural interest groups have perceived the legislation as a socialist attack upon their role as custodians of rural land. It is true that the measure was hurriedly introduced and without time for sufficient consultation with the interest groups involved in the industry (see Northfield 1979:223).

The tenor of such antagonisms were well exposed in the evidence presented to the Northfield enquiry. The Agricultural Law Association expressed the view that:

> the result of this new right of succession will be that few, if any, individual landowners will be willing to grant new tenancies in the future.

The National Federation of Young Farmers' Clubs argued:

> the legislation making it likely that a farm will be out of the landlords hands for three generations is now responsible for a decrease in the number of tenancies available. This legislation does great damage to the majority.

The Northfield Committee (1979) concluded that it was yet too early to judge the success or otherwise of the Act, but that modifications were necessary in order to increase the number of tenancies available. Their proposals included i) setting a maximum area of land above which successors would not have the right to succeed to the whole tenancy; ii) stricter rules on the eligibility and suitability of successors; iii) encouragement of early transfers iv) the provision of some lettings without the security of tenure and v) the encouragement of landlord-tenant partnerships.

Such a mixture of proposals, while reflecting the divergent views of members of the committee (see Munton 1984), also highlighted a central question concerning the maintenance of existing landlord-tenant relations: how much security of tenure should tenants sacrifice to make letting land a more attractive proposition and therefore recreating more tenancies? Onus was thus moved from landlords and towards their tenants in devising schemes to regenerate the let sector. From the position of the landlord, and indeed the majority of related interest groups, a reduction in the security of tenure was a legitimate means of increasing the number of tenancies. The original justification for the 1976 Act - enabling continuity of the farm family for the tenants - was increasingly

regarded as an inconsequential feature of a piece of poor
legislation.

The return of a Conservative Government in 1979 and its
re-election in 1983 with an increased majority provided the
political context in which to challenge the basic elements of the
1976 legislation. A new tenancy bill was prepared, largely based
upon a set of joint proposals from the NFU and the CLA aimed at
reducing the security of tenure to the lifetime of existing tenants.
Any farmer who held the right to pass on his farm under the 1976
legislation was to retain his succession under that Act, but all new
tenants were to forego succession with the letting ending upon the
tenants' death. As well as abolishing the statutory succession
scheme established under the 1976 legislation, important amendments
were also made to the rent arbitration formula under section 8 of
the Agricultural Holdings Act of 1948. Prior to the 1984 Bill, farm
rents were determined by i) the terms of the tenancy - for example,
rents are lower where the tenant is responsible for repairs;
ii) what the rent would be if the landlord let the farm to a brand
new tenant on the open market and iii) landlords' improvements and
potential rateable values. Under the new proposals, supported by the
Tenants Farmers' Association, it was argued that arbitrations must
make allowances for the distorting effects that the lack of
tenancies was having on the bidding-up of rent levels. As a
consequence, the new legislation sought to modify the criteria on
which rent levels could be established. The 1984 Act thus stipulates
that:

> The rent properly payable in respect of a holding shall be the
> rent at which the holding might reasonably be expected to be
> let by a prudent and willing landlord to a prudent and willing
> tenant taking into account all relevant factors, including (in
> every case) the terms of the tenancy (including those relating
> to rents), the character and situation of the holding
> (including the locality in which it is situated), the
> productive capacity of the holding and its related earning
> capacity, and the current level of rents for comparable
> lettings.

The last factor - 'current level of rents for comparable
lettings' - was now to be adjusted to discount the effect that the
scarcity of holdings had on existing rent levels. Thus, the

arbitrator should disregard:

> any element of the rents in question which is due to an
> appreciable scarcity of comparable holdings available for
> letting on such terms compared with the number of persons
> seeking to become tenants on such farms.

The attention given to this issue has been viewed as a concession to
the farming lobby in return for the abolition of statutory
succession rights, and thus very much as the NFU's and Tenant
Farmers' Association's bonus in the compromise NFU-CLA 'package' of
proposals which laid the basis for the Act. The ways in which the
'scarcity value' is to be consistently measured and implemented is,
however, vague. Indeed, as one commentator has suggested:

> He would be a bold man who would try to forecast what
> difference the new rent formula may have, and in this respect
> the Act was merely clothing the present practice with legality.
> It may thus be difficult to trace any causal connection between
> the changes in the rent formula and future arbitrated rent
> levels. This does not necessarily mean that the changes are
> totally devoid of effect. If they increase the confidence of
> tenant farmers in the arbitration system for rents they will
> have served a useful purpose.
>
> (Muir Watt 1984:85).

The passage and enactment of the bill provided the focus for
considerable debate among the related interest groups. Unlike the
debates associated with the 1976 Act which emphasised the continuity
of family tenancies however, emphasis has been placed upon
establishing a more 'efficient' and flourishing landlord-tenant
system. The newly-formed Tenants Farmers' Association, established
to protect tenants' interests, put forward its own package of
proposals in 1982. The package included the abolition of succession
for new tenants, rents based on the productive capacity of the land,
tax incentives for landlords to let, tenant retirement at 65, and
tax disincentives for occupation of more than 2000 acres. Hence,
from the outset, despite their need to distinguish themselves from
the NFU and the CLA, their policy objectives were largely consistent
with the more established interest groups. The Association sees
itself as a ginger group committed to speaking out for the private
landlord-tenant system and has been anxious to avoid giving the
impression that it has aspirations to be a militant tenants' union.
A major practical function that the TFA performs (and, it argues, a

major reason for its increasing membership), concerns its advisory
service to tenants, assisting its members during rent review
procedures. While the now three-sided agricultural-landowning lobby
developed a consistency in their arguments relating particularly to
the abolition of statutory succession, the Labour Party was critical
of the measure. They agreed that the move to bring rent fixing
criteria in line with the earning potential of the land was a
constructive element of the 1984 legislation but continued to argue
in support of statutory succession in order to maintain family
continuity. Moreover, the Farmers' Union of Wales consistently
opposed the legislation, maintaining their claim that three
generational succession should remain intact. They also canvassed
for special consideration to be given to 'family farming' areas
where succession was seen as more crucial for the maintenance of
existing farm structures. The long awaited agreement between the
more powerful NFU and the CLA, however, and the apparent eagerness
of Government to enact their proposals, were crucial elements in the
successful passage of the legislation.

Despite the general impression created that the Act's main
purpose was to 'tidy-up' existing legislation and to streamline the
position of landlords and tenants, it could have serious
repercussions for tenants' security in the future. Representatives
from the land agents Strutt and Parker have argued, for instance:

> Some have called this Act an irrelevance - a mouse - but we see
> it as an important turning point from the post-war attitude
> which provided over-security and over-fair rents to tenants...
> This Act is a first step in the right direction towards more
> farm tenancies. It should help to create a fairer balance
> between landlord and tenant.

The major stated objective of the Act - to produce more farm
tenancies and to encourage 'new blood' into the farming sector - has
concentrated upon modifying the relations between landlords and
tenants. This has the additional consequence of allowing the owners
of land more control over their property. Current trends in the land
market suggest that the expansion of in-hand farming, by landlords
and institutions as well as increased owner-occupation are just as
responsible for producing the scarcity of tenanted land as is the
reluctance of landowners to create or sustain tenancies. In this

respect state legislation based around a predominant ideology of the 'farming ladder' as opposed to the continuity of the family farm has focused upon only one progressively contracting sector of the land market in an attempt to boost the number of vacant tenancies. One of the Act's most sternest critics, Lord Northfield, has dubbed it 'counter-productive', questioning whether new tenancies will result. As he argued in the House of Lords 'how far can we go in preserving islands of individual landowners for what is going to be a very marginal increase in tenancies'.

CONCLUSION

The abolition of statutory succession for tenants in 1984, which has eventually 'clawed back' landlords' control over their land, together with the increased relief given in the amendments to Capital Transfer Tax in 1986, are examples of state action to reinforce the autonomy and control of landowners. Similarly, Goodchild and Munton (1985) have shown that the abolition of Development Land Tax will particularly benefit landowners who consider selling off land for urban development. In the 1980s, the vehemence of political attacks upon landowning groups has waned with the re-emergence of a strong utilitarian ideology supported by the practice of deregulation (see Denman 1980). The change in the complexion of Government since 1979 and particularly the rise of New Right ideology has provided the context for the particular changes relating to agricultural tenure legislation outlined in the final part of this paper. The CLA has been particularly successful in presenting the private landlord as the Guardian of the Countryside, and advocating within this political context policies of self-regulation (Lowe et al 1986). What is perhaps of equal significance is the extent to which other interest groups such as the NFU and the Tenant Farmers' Association have also supported the ideological shift of emphasis despite problems of unifying sectoral interests within their organisations. Increasing pressures, however, are now being placed upon the dominant ideologies of production, self-regulation and the custodial role of the private landowner, not least because of the difficulties of legitimising these at a time of general economic recession, over-production of agricultural goods,

large-scale state subsidisation and a growing realisation of the
environmental costs.

REFERENCES

Abercrombie, P. 1930. The English countryside. In The political
     quarterly in the thirties, W. A. Robson (ed). London: Allen
     Lane.
Bornstein, S., Held, D. and Krieger, J. 1984. The state in
     capitalist Europe. London: George Allen and Unwin.
Denman, D. 1980. Land in a free society. London: Centre for Policy
     Studies.
Goodchild, R. and Munton, R. J. C. 1985. Development and the
     landowner: an analysis of the British experience. London:
     George Allen and Unwin.
Harrison, A. 1982. Factors influencing ownership, tenancy, mobility
     and the use of farmland in the United Kingdom. Commission of
     the European Communities: Information on Agriculture No 74.
Jones Lang Wootton, 1983. The agricultural land market in Great
     Britain. London.
Laski, H. H. 1934. Liberty and the modern state. London: George
     Allen and Unwin.
Locke, J. 1960. Second treatise on government. Cambridge: Cambridge
     University Press.
Lowe, P., Cox, G., MacEwen, M., O'Riordan, T. and Winter, M. 1986.
     Countryside conflicts: the politics of farming, forestry and
     conservation. Aldershot: Gower.
Massey, D. and Catalano, A. 1978. Capital and land: landownership by
     capital in Great Britain. London: Edward Arnold.
Macpherson, C. B. 1973. Democratic theory: essays in retrieval.
     Oxford: Clarendon Press.
Munton, R. J. C. 1977. Financial institutions: their ownership of
     agricultural land in Great Britain. Area 9, 29-37.
Munton, R. J. C. 1984. The politics of rural landownership:
     institutional investors and the Northfield enquiry. In Rurality
     and locality, P. Lowe and T. Bradley (eds), 167-178. Norwich:
     Geo Books.
Munton, R. J. C. 1985. Investment in British agriculture by
     financial institutions. Sociologia Ruralis, 25, 155-173.
Muir Watt, J. 1984. Agricultural Holdings Act 1984, 1. Estates
     Gazette, 271.
Newby, H., Bell, C., Rose, D. and Saunders, P. 1984. Property,
     paternalism and power: class and control in rural England.
     London: Hutchinson.
Northfield Committee Report, 1979. Report of the Committee of
     Inquiry into the acquisition and occupancy of agricultural
     land. Cmnd 7599. London: HMSO.
Offe, C. 1972. Political authority and class structure.
     International Journal of Sociology, 2.
Roger Tym and Partners, 1983. Agricultural performance analysis.
     London.
Sturmey, S. G. 1955. Owner farming in England and Wales. Manchester
     School, 23, 245-268.
Thompson, F. M. L. 1963. English landed society in the nineteenth
     century. London: Routledge and Kegan Paul.
Thompson, F. M. L. 1979. Cited in Northfield Committee Report,
     Report of the Committee of Inquiry into the acquisition and
     occupancy of agricultural land. Cmnd 7599. London: HMSO.

Trail, B. 1981. Land values and rents. University of Manchester, Dept. of Agricultural Economics: Bulletin 275.

Veblen, T. 1923. Absentee ownership and business enterprise in modern times. New York: Sentry Press.

9 Investment styles and countryside change in lowland England

CLIVE POTTER

INTRODUCTION

Despite an increasingly persuasive body of evidence describing the
pattern of post-war countryside change, there is still remarkably
little research that offers a convincing explanation of the process
of change at the micro-level. The contemporary debate surrounding
agriculture and the environment is fuelled by an intense public
reaction to landscape and habitat loss which has both a personal and
political dimension. This inevitably means that the discourse is
rather better informed about symptoms than it is knowledgeable about
causes. The development of a more diagnostic account has yet to be
made. Certainly the statistical evidence points towards
environmental externalities as an endemic feature of an
industrialised agriculture. As Heady (1975:5) argues: 'technological
and structural changes in agriculture, when implemented by a large
enough group of farmers, nearly always involve complex
externalities'. The Nature Conservancy Council's (NCC) documentation
of, inter alia, a 30-50 per cent loss of ancient lowland woods, a
40 per cent loss of lowland heaths, a 75 per cent loss of hay
meadows and a 50 per cent reduction in fens and valleys and basin
mires since the Second World War, provides gloomy corroboration of
Heady's claim. Commenting on these trends, the Council marks a
rapidly approaching situation in which 'the designated areas (SSSIs,
nature reserves, National Parks,.....) are becoming oases in a
wildlife desert' (1984:18).

There is much less agreement at the level of explanation. Within the debate itself different attributions take on political significance as interest groups come to recognise that diagnosis leads powerfully to policy prescriptions. In this respect it is tempting to follow the farm lobby by opting for what can be described as an idiosyncratic interpretation, in which farming 'mavericks' (Cox and Lowe 1983:65) are incriminated for any changes to the conservation estate that might occur. By emphasising the seemingly gratuitous action of identifiable individuals, the lobby is able to confine its prescriptions to matters of conscience rather than problems of policy.

The difficulty with this interpretation is an implied ignorance of the very characteristic pervasiveness of habitat loss - which the NCC figures barely hint at. Mavericks no doubt exist, but hardly in sufficient numbers or spread to explain the extent of damaging change.

What might be a logically opposed explanation is offered by those who take a 'policy determinist' approach, emphasising the role of larger structural influences in determining the pace and direction of countryside change. The impression given by writers such as Body (1983, 1984) and Bowers and Cheshire (1983) is of a powerful policy-machine exerting an irresistible influence. On this reading, farmers carry out damaging operations because market intervention and production grants make them irresistably attractive. Motivation and action become disconnected as the inducement to further, potentially damaging operations over-rides any personal inclination for restraint. In the words of the NCC (1984:3): 'farmers are inextricably bound in a web of financial structure.....and few have the means to operate outside it'.

Analytically, this represents a considerable advance on the 'black sheep' analogy. It acknowledges that conservation damage and loss may be more often incidental to larger moving forces which are structural in origin, rather than gratuitously imposed by individual whim. Recognising this enables debate to move forward to consider the radical opportunities offered by agricultural policy itself. The problem is no longer seen as resolvable solely by the farmer. There are problems even with this approach, however. To begin with, we have only a vague idea of the behavioural impact of policy instruments at farm level, and especially of the role played by advice, tax reliefs and the grant system in fashioning investment and management decisions. Thus, while in the case of grant aid for what are damaging operations, it can be said that public monies <u>fund</u> an externality-generating activity, it cannot be said that receipt of these state aids everywhere and always <u>determine</u> those decisions. There is often a 'non-additionality' problem here in that investment incentives may not generate new capital spending, but merely subsidise investment which would take place anyway. There is, moreover, the ambiguous nature of the supply response in a protected market and the extent to which technological change is an endogenous force for change. Capstick (1983:263), for instance, believes that: 'whatever the mix and pattern of exogenous forces pressing on the industry, they have never been such as materially to deflect the industry from its upward course'.

This approach also denies any significant role for human agency, just as the previous view over-emphasised it. By assuming that farmers are merely reactive 'policy dopes', it risks treating causal influences as separable when they are interactive. The Bowers and Cheshire model is a case in point. In this, the influence of price support, mediated through inflated land values, is given causal priority in a way which 'intensification' is apparently sufficiently explained by movements in these economic variables alone. This leaves out of account those deep-set sociological factors which determine adoption of technological change and the role of 'family processes' (to use Marsden's (1984) phrase) in explaining investment and management decisions which may be environmentally damaging. An instance of comprehensive improvement which takes place following farm expansion might therefore be explained by a farmer's reaction to the inflated purchase price, but it is at least as plausible to talk of investment decisions undertaken to solve problems of family development and succession. A policy determinist approach also discounts farmers' own rationalisations of past investment decision. It is usual to caution against an over-reliance on answers to Why? questions, though as

Marsh (1983:107) comments, 'actors do know a great deal about why they behave as they do, and we must not eschew the insights they have'.

The point, then, is not that any of these explanations is wrong, but that none of them is sufficient, because the nature of interactions between various layers of influences have still to be worked out. A good analogy is to imagine attempting to describe the contents of a room by looking through two or three small windows set on different walls. Each view would be different and without entering the room we can never confirm the accuracy or completeness of our description. As David Harvey (1983:2) puts it: 'the view from any window is flat and lacks perspective. When we move to another window we can see things that were formerly hidden from view.....by moving from window to window and carefully examining what we see, we come closer and closer to understanding.....'. In terms of this analogy we can see that there are three possible 'windows' on countryside change.

First, we can identify a macro perspective based on a general model of policy and technological cause and effect. Factors such as price-support, inflated land values and investment incentives, as well as the pressures for change exerted by indebtedness and the treadmill of technological change are all pertinent here in providing a post-hoc rationalisation of the direction and pace of change. Secondly, there is an intermediate perspective which focusses on various pressures for change which arise periodically within the family farm. This perspective might point, for instance, to the environmental damage contingent on management and investment changes following succession. It can be used to explain episodes of countryside change. Finally, change can be studied at the level of farmers' personalities, emphasising achievement motives. Environmental damage may result from behaviour which is conditioned by investment norms and notions of what makes for good practice. An adequate understanding is likely to demand the insights provided by each of these levels of emphasis.

ULTIMATE AND PROXIMATE EXPLANATIONS

All this suggests that interaction between agents and structure is
so continuous it is inappropriate to attempt an analysis in terms of
a one-way causal relationship. A farmer's business world is, in
Giddens' terms, 'bounded by unacknowledged conditions of action on
the one side, and unintended consequences of action on the other'
(Giddens, 1981:7). We can encapsulate both the conditioned nature of
much farmer behaviour (in terms of factors which may not be
accessible to the agent) and the unintended aspects of countryside
change, whilst also presenting farmers as knowledgeable subjects.
This was the theoretical position which underpinned research
conducted by the writer in three lowland counties - Suffolk, Norfolk
and Shropshire - in 1983/84. Some 175 randomly sampled farmers were
interviewed. The main methodological requirement was that the sample
should make possible the study of activity patterns rather than
'unique life situations'. As Brenner (1980:35) has said with respect
to his theory of action: 'To understand an action, it is necessary
to understand the situation of its occurrence, and also to identify
the temporal and hierarchical structure.....and all must be
investigated from the active agency perspective of the parties
involved'. Since net post-war losses of habitat and landscape
quality are the product of two contervailing activity patterns -
those leading to further land use intensification against those
promoting restraint, reconstruction and/or benign neglect - two
representative forms were studied: conventional investment in land
improvement (which is capital spending on things such as drainage,
woodland clearance, hedge removal and grassland improvement) and the
range and quality of conservation practice.

Let us examine the more dominant, regular and consistent of
these forces first.

Land improvement is a catch-all term for many farm operations
which may result in a degradation or loss of environmental quality.
It is defined here as any durable realisation of the productive
potential of the soil due to capital investment, the irregular
expenditure of labour or the commitment of fixed equipment.
Motivation towards such action was found to be strongly expressed

throughout the sample, and though such motivation appears
voluntaristic, it was also conditioned and reinforced in subtle
ways.

The 'achievement motive' was particularly apparent. In support
of Wolpert (1964:545), it appears true of the farmer that
'aspiration levels tend to adjust to the attainable, to past
achievement levels, and to the levels achieved by other individuals
with whom he compares himself'. The effect of technological change
and policy inducement is to bring land improvement within reach, to
make it attainable. In this sense, the achievement motive is both a
cause and effect, since the extent to which values are held is
susceptible to the distribution of opportunities and constraints
throughout the system.

> It gives me a sense of progress. Farming these days is so much
> a matter of staying on the same roundabout. It's good to be
> able to make an improvement in some part of the job.
>
> (Suffolk farmer, 390 acres).

> A lot of the business of farming is out of my hands. Plant
> breeders and spray manufacturers can be relied on to do all the
> trial and error work. Land improvements, though, is something I
> can do myself with little risk or worry.
>
> (Shropshire farmer, 250 acres).

On the other hand, improvement implies change. It was found,
however, that even older respondents could reconcile the business of
improvement with a latent desire for the status quo by invoking the
higher desideratum of orderliness. As Denman (1957:117) has written:
'improvement means change, a giving of new for old, a planned
purpose: sometimes it is deeply surgical. The sentiment that urges
forward improvement and comes nearest to love of the land is a
craving for orderliness. (Farmers) improve their land because they
cannot brook neglect or untidiness'. Being 'up to the teeth', as one
Shropshire farmer graphically put it, is obviously a matter of
professional pride.

Clearly this feature lends support to the thesis that
environmental change is often the 'unintended consequence' of innate
patterns of action and decision. More than that, it derives from a

deeply ingrained behavioural regularity which gives an dynamic to
the process of change.

There was a majority involvement in the practice of
improvement, matched by a generally acute perception of outstanding
investment need. Something like 40 per cent of the sample had
further improvement planned, with varying perception of outstanding
need according to circumstance and length of establishment.
Investment level was measured through farmers' estimates of physical
change to the 'utilised agricultural area' which, despite its
shortcomings as a method of enumeration, allowed for a comparative
analysis of land improvement undertaken by farms of different size,
type, occupance and location. About one-third of respondents
recorded no change in this variable since 1970, with varying degrees
of change elsewhere.

From detailed case study work it further proved possible to
distinguish different 'investment styles' or ways of proceeding. On
the basis that commitment to the idea of land improvement is
widespread, differences in the timing, rate and extent of
improvement express a constant adaptation to opportunities and
constraints by the farmer (see Fig. 1). In this sense, 'style'
becomes a 'proximate' explanation of countryside change. It can be
used to predict the character of any countryside change which
arises. The character of environmental damage and loss is related to
how episodic or continuous it is and the extent to which it is
concentrated or spread between holdings in a locality. These are
significant factors where policy reform is concerned (see below).

Briefly, two principal styles can be distinguished:

1. A systematic style exhibited by well-established and capital rich
farmers, whose capital investment is triggered or 'enabled' by
discrete changes of circumstance such as new land purchase, changes
to or from tenancy, succession of a son and/or enterprise change and
expansion. Land improvement here takes on a programmatic aspect and
is linked to farm business growth and development, itself the
expression of diffuse and complicated changes in the policy and
business environment. Environmental damage and loss is likely to be
highly visible and controversial.

2. An incremental or problem-solving style exhibited by farmers constrained by a variety of financial and/or situational factors such as an inability to generate sufficient investment funds due to low credit worthiness/low margins; absence or uncertainty of succession, vulnerability to market fluctuations due to a sub-optimal enterprise mix, etc. Land improvement is much less visible because it is small scale and continuous, though it may nevertheless be cumulatively damaging. Incrementalists will be highly risk-averse and investment is dictated by outstanding need.

The basis of this characterisation is shown in Table 1. Clearly, it is an intuitively derived set of distinctions which need further refinement and empirical validation. However, by creating a category of accessible, foreground variables, the way is clear for a fresh approach to research on countryside change.

Table 1 Investment Styles

	Systematic	Incremental
planning horizon	long-term, initiatory	short-term, reactive
historical level of provision	high	low
reasons for improvement	linked to business grow/development	linked to perceived need
sources of advice	external, official, commercial	internal, informal
grant aid receipts/ scheme	high	low
use of manpower/ equipment	contractor services/ hired labour	farmer's own labour
nature of individual schemes	extensive and comprehensive	small and ad hoc
nature of environmental impact	highly visible and controversial: 'black sheep' activity	less visible and cumulative: 'background noise'

CONSERVATION PRACTICE: SHALLOW OR DEEP?

Turning now to conservation activity, there is a very evident difference in motivation and action here. The gap is a qualitative one and owes something to the discrepancies in implementation

success at policy level (see Potter, 1984). The survey revealed a
pattern of investment much more irregular, variable in quality and
fleeting in duration, than conventional forms. A number of
fashioning influences were at work. Some of these, which I have
called 'introspective' influences, were highly personal and
idiosyncratic, whilst others, which might be labelled 'extrovert'
influences, operated at group level. To facilitate study of these
introspective and extrovert influences, some 40 members of the
Farming and Wildlife Advisory Group (FWAG) in Shropshire and Suffolk
were interviewed.

It is basic to the farm lobby's advocacy of the 'voluntary
approach' that the agriculture-conservation problem is seen as
self-correcting. Investment in conservation which is already
occurring, so the argument goes, has only to spread to a wider
constituency by permissive means for the difficulty to be
satisfactorily resolved. As an emblem of this 'stewardship ideology'
(Cox et al, 1985), FWAG is obviously significant. In view of such
parallel work, it was of interest to discover that stewardship
still existed as a category of farmers' consciousness and to assess
the quality of provision at farm level. It is plain that FWAG is
both a transmission channel for an ideology and a practical,
problem-solving tool.

An alternative but not mutually inconsistent interpretation
would see FWAG as constituting small homogenous groups acting to
produce optimal amounts of a collective good (conservation). This
fits well with the introspective yet also 'extrovert' nature of
conservation activity. Olson's (1965) theory of public goods
predicts that individual willingness to contribute to the group's
provision of such goods depends on: (a) the noticeability of
members' behaviour and (b) the perceived effectiveness of their
behaviour in contributing towards or detracting from that provision.
The survey provides evidence of the operation of both influences on
the level of conservation provision. The noticeability component,
for instance, was both a cause of restraint and a reason for
positively expressing good conservation practice. This is because an
individual can experience negative group sanctions in the form of
the disapproval of fellow-farmers by noticeably not contributing -

for instance, by visibly destroying habitat and landscape - as well
as positive rewards, and in this connection the annual FWAG
competition is important as an opportunity to demonstrate active
involvement in conservation.

> There's a tremendous sense of disapproval from fellow farmers
> if someone goes too far.....something that didn't happen 10
> years ago.

> I can see a time when all farmers will find themselves under
> pressure from beady-eyed neighbours on the conservation front.

A contradiction between each of these forces was also evident,
however, since in some cases highly visible enhancement
conservation - tree planting, the creation of new features and other
ameliorative work - only became desirable once planned improvement
programmes had been executed.

> Conservation is all part of a well-run business.....but it has
> to be organised as you want it first.....otherwise it's a
> millstone round your neck in 10 year's time.

> You have to be in control before you can see where conservation
> is needed.....I see it as a very deliberate process. The final
> piece of the jigsaw.

The impression built up from detailed case studies is of a
'balanced configuration' of attitudes. Many FWAG farmers, it seems,
achieve compromise by segregating the two activities in time and
space. Segregation in time means that conservation enhancement takes
place on farms with the most 'complete' investment histories.

Regarding the perceived effectiveness of contribution, many
FWAG members saw themselves as opinion leaders or early innovators.
Some indicated that their contribution was an effective and
influential one because of its exemplary character. Certainly a
process of diffusion from farmer to farmer can be identified, though
the characteristic complexity and divisibility of conservation
practice suggests a tendency for quality of provision to decline
through time and space. In view of the segregation phenomenon, it
may be that a visible diffusion of conservation activity is nothing
more than increasing numbers of farmers reaching a point of
completion or near-completion in conventional investment programmes.
The conservation implications of this can only be guessed at. It

reinforces the need for a careful distinction to be made between
enhancement and protection conservation, and suggests a basic flaw
in the voluntary approach.

THE POLICY IMPLICATIONS

The lessons of this analysis of countryside change as process are
generally pessimistic although challenging in specific ways. To
begin with, the balance of motivation and action at farm level for a
majority of those sampled gives scant support to a view of the
agriculture-conservation problem as self-correcting. We have seen
how, for instance, conventional and innovative (conservation)
behavioural patterns are qualitively different in terms of the
policy-determined reward structure, the depth and generality of
farmer commitment, and the regularity and consistency of action. The
prospect on balance is for a continued loss and deterioration in the
aesthetic and ecological quality of the farmed landscape, unless
policy objectives can be strategically set at a level beyond that of
the individual farm.

Methodologically speaking, it appears difficult to establish
causal links between policy and behavioural variables with any
degree of certainty. Rather, there is a need for study of movements
in various proximate variables such as investment style. But again,
we must remember that this is itself a form of shorthand for
recombinations among ultimate variables. Among the most decisive
likely future 'movements' will be the evolution of investment style
within already extant farms and the movement of land between farmers
practising different styles. By focussing on these the analyst comes
close to capturing the dynamic of countryside change, while also
being free of the entanglements of contingency and circumstance. It
is an especially useful way of ordering our ideas about the impact
of possible CAP reform on the environment.

From the environmental angle it might seem that the much mooted
reductions in levels of farm support will effectively 'bracket' the
problems of habitat and landscape change, assuming we are correct in
thinking that historically high levels of spending have been prime
movers in the process of change to begin with. Unfortunately, our

alternative approach throws doubt on the possibility of such a happy symmetry by demonstrating that the past is a very unreliable guide to the future. This is because countryside change is on a trajectory, driven by continuing structural change due to family pressures and rigid subjective assessments of what makes for good farming. Both will be slow to alter in response to policy signals - the farmer cannot be pulled as easily as he was pushed (see CPRE/IEEP, 1985, for discussion).

The high valuation of land improvement suggests, for instance, a propensity merely to allocate funds away from other investment items to sustain planned programmes. There is, moreover, scope for individuals to continue to adopt their investment styles rather than adjust their levels of investment activity - for instance, to 'step down' from a systematic to an incremental way of proceeding. In this respect it is not the quantity but rather the quality and pattern of capital spending which responds. A shift of this sort would make change more difficult to detect over particular areas of land and the use of 'safety net' controls less reliable. Indeed, the fact that an investment style is specific to whole businesses means that partial development control might bypass whole farms where incrementalism was practised, as well as single in-field and small scale operations. The systematic nature of such neglect would surely be problematic though by the same token, of course, the argument for a general order-making power linked to notification is strengthened.

The impact of retrenchment on land mobility brings us to the question of repercussions between farms. If, as expected, the rate of amalgamation accelerates during a squeeze, the effect could be to short-circuit the process of change by moving more land into the hands of those least vulnerable farmers (because of lower levels of bank borrowing, length of establishment, and so forth) ready to treat newly-acquired land as an enablement for further improvement. In these circumstances, retrenchment changes the overall composition of investment styles by pushing the 'natural improvers' to the fore and increasing their absolute and relative influence on the rate of countryside change.

In view of these and other reactions, retrenchment of farm

spending seems unlikely to be an unalloyed environmental 'good'. At best the impact must be indeterminate and repercussive; at worst actually detrimental. It is important that this possible asymmetry in the farm response to, respectively, rising and declining real price support is recognised.

Similarly, a permissive 'voluntary' conservation policy must be severely limited in scope while the implementation gap remains. With the emphasis on reconstruction of existing policy instruments, the challenge to policy-makers is to recognise these complexities and institute in response a system of controls, state aids and subsidies which aim to promote farm income, conservation and social objectives simultaneously. The notion of whole farm funding and control is crucial to success here. This concept recognises the need to place restraint within its business context (see above) and of the potential attached to income support as a means of conservation as well as an end of farm support (Potter, 1983). Policy itself may have to become discriminating if the polarising trends in farm structures and investment styles are to be accommodated. There is a suggestion that while the 'stick and carrot' approach - partial controls combined with conservation incentives - may work for 'enabled' farmers, it will be largely ineffective for those constrained businesses responsible for the significant 'background noise' of countryside change. Substantive income support, linked to conservation requirements, is a feasible option here.

This paper has shown that, in order to understand the full complexity of contemporary countryside change, we need to accept that different aspects of the process can be understood by referring to factors operating at farmer, family and policy levels respectively. Generalisations can still be made, however, since typical patterns of behaviour - themselves adaptations to opportunities and constraints as they exist in the system - give rise to particular types of environmental damage and change and analysis which can specify such relationships provides insights of considerable relevance to policy reform.

REFERENCES

Blacksell, M. and Gilg, A. 1981. Countryside planning and change.
    London: George Allen and Unwin.
Body, R. 1982. Agriculture: the triumph and the shame. London:
    Maurice Temple Smith.
Body, R. 1984. Farming in the clouds. London: Maurice Temple Smith.
Bowers, J. K. and Cheshire, P.C. 1983. Agriculture, the countryside
    and land use. London: Methuen.
Brenner, I. 1980. The structure of action. Oxford: Basil Blackwell.
Capstick, C. W. 1983. Agricultural policy issues and economic
    analyses. J. Agric. Econ. 34, 263-377.
Centre for Agricultural Strategy, 1978. Capital for agriculture.
    Reading: CAS Report No 3.
CPRE/IEEP, 1985. Environmental implications of future CAP price
    policies. London: Proceedings of seminar, CPRE/IEEP.
Coleman, D. 1984. Economic pressures on the environment. In
    Investing in rural harmony: a critique. Reading: CAS Paper
    No 16.
Cox, G. and Lowe, P. 1983. A battle not the war: the politics of the
    Wildlife and Countryside Act. Countryside Planning Yearbook, 4,
    48-76.
Cox, G., Lowe, P. and Winter, M. 1985. Land use conflict after the
    Wildlife and Countryside Act 1981: the role of the Farming and
    Wildlife Advisory Group. J. Rur. Stud. 1, 173-183.
Denman, D. R. 1957. Estate capital: the contribution of
    landownership to agricultural finance. London: George Allen and
    Unwin.
European Commission, 1984. Com. 83 (559), Proposals for new
    structures directives, Brussels.
Giddens, A. 1981. A contemporary critique of historical materialism.
    London: Methuen.
Harvey, D. 1983. The limits to capital. Oxford: Basil Blackwell.
Heady, E. 1976. Externalities in the transformation of agriculture.
    Iowa, USA: Iowa State University.
Lords Select Committee on the European Communities, Agriculture and
    the environment, 20th Report. London: HMSO.
Marsden, T. 1984. Capitalist farming and the farm family: a case
    study. Sociology. 18, 205-224.
Marsh, C. 1983. The survey method. London: George Allen and Unwin.
Nature Conservancy Council, 1984. Tenth annual report. London: HMSO.
Olson, M. 1965. The logic of collective action. Cambridge, Mass:
    Harvard University Press.
Opp, K. D. 1982. The evolutionary emergence of norms. Brit. J.
    Social Psyc. 21, 139-149.
O'Riordan, T. 1983. Putting trust in the countryside. In A
    conservation and development programme for the UK. London:
    Kogan Page.
Potter, C. A. 1983. Investing in rural harmony: an alternative
    package of agricultural subsidies and incentives. Godalming:
    World Wildlife Fund.
Potter, C. A. 1984. Investing in rural harmony. In Investing in
    rural harmony: a critique. Reading: CAS Paper No 16.
Wolpert, J. 1965. The decision process in a spatial context. Annals
    Assoc. Am. Geog. 54, 159-162.

10 British agriculture under attack

GEORGE PETERS

INTRODUCTION: CHANGES IN AGRICULTURAL POLICY

In the past forty years Britain's agricultural policy makers have
seldom been free of critics. But the chorus of protest has become
increasingly vociferous since accession to the European Community in
1973 and, during the present decade in particular, few opportunities
have been missed to deride the system which is now in operation. The
quietest period, in contrast was between the passing of the 1947
Agriculture Act and the introduction of the Common Agricultural
Policy (CAP) and it is worthwhile to begin by accounting for this.

Traditional policy was based on deficiency payments and direct
production grants. Although deficiency payments were by no means
easy to administer their economic effect was to prevent a wedge of
distrust developing between farmer and consumer. After decontrol in
the early 1950s the farmer was able to sell produce in the market at
first hand prices determined principally by the cost of imports,
many of which could enter without major restriction. Records were
then kept of sales and average prices and the difference was made up
to guaranteed levels by Exchequer payments. Details varied,
especially in the important case of milk, but the end result was a
policy framework which appeared defensible. Consumer food prices, in
particular, were maintained on the base of world prices and
availability. Furthermore, since taxation was progressive it could
be argued that the better off made contributions - partially offset
for them by the low food prices - which were of special benefit to

poorer groups in society. Overseas suppliers were damaged, but by
less than if domestic support had relied on frontier protection,
while farmers operated with a measure of certainty which enabled
them to plan effectively.

The system was not uniformly applauded however. One great
disadvantage was that the guarantees were 'open ended' so that
production increases, or a decline in import prices, would force up
the Exchequer cost. There was disquiet among economists, for
instance, who argued that the balance of payments case which was
often quoted in defence of support was a weak one and some political
scientists were uneasy. In their famous study The State and the
Farmer, published in 1962, Self and Storing complained about the
power of the farming lobby and the cosiness of the relationship
between the Ministry and the National Farmers' Union which was
fostered by the annual review process initiated with the 1947
Agriculture Act, and reinforced by the 1957 Act, which set limits
to any decrease in aggregate guarantees and to the individual
product prices which could be imposed. Although it did have to be
amended by the introduction of standard quantities, minimum import
prices and agreements on supply from third countries, the system was
maintained until the Conservative government, elected in 1970, began
to alter policy to rely more on frontier protection. This situation
was in turn overtaken by eventual accession to the EEC.

What then became important was that protection through tariffs
and levies could cause a rift between farmer and consumer since
'low' world prices would no longer determine the price of food. In
one sense, of course, the changeover came at a fortunate time for
the government since in the early 1970s there was a temporary but
distinct rise in international market prices which could be
presented as a development which would have raised prices in any
case. Indeed, supporters of the EEC in the 1975 referendum campaign
made much of this. Since then, however, criticism has grown. It has
been fuelled by a number of factors: notably the constant complaint
about EEC financing and the British contribution, the scandal of
intervention purchases and, especially of late, the feeling that
agriculture is increasingly pampered and protected and is uniquely

favoured at a time when the economy at large has experienced traumas
hitherto unprecedented in the post-war period.

Although the effects of the CAP on Britain have been widely
analysed and frequently deplored by academic economists the
publicity pool has recently been scooped by the Conservative MP and
farmer Richard Body whose two books Agriculture: The Triumph and the
Shame and Farming in the Clouds have had tremendous impact. It is
important to appreciate that Body regards himself as an ally of
farmers, although many in the farming community might dispute this.
The other attack is of a rather different nature. While many argue
that farmers are unduly privileged other writers picture them as
despoilers of the land. Farming is presented as the generator of
externalities which are seen as harmful to the community at large,
or perhaps more strictly to the interests of other pressure groups
within it. Such arguments have been associated with the rise of
environmental awareness and the result, to all intents and purposes,
has been an alliance between those who have long complained about
the apparently specious general economic arguments for agricultural
support and those who stress its deleterious environmental impact.
It would be superfluous to provide a catalogue and there is no need
to do more than quote a few lines from the Introduction to Marion
Shoard's book The Theft of the Countryside, published in 1980, since
it provided one of the more outspoken critiques:

> Our empire may have passed away; our industrial strength may be
> tottering; but the matchless charm of our countryside - that
> was ours before imperial power or economic hegemony, and it
> survived them. Until now. Although few people realise it, the
> English landscape is under sentence of death. Indeed, the
> sentence is already being carried out. The executioner is not
> the industrialist or the property speculator, whose activities
> have touched only the fringes of our countryside. Instead it is
> the figure traditionally viewed as the custodian of the rural
> scene - the farmer.
>
> (Shoard 1980:9).

She goes on to speak of destruction of hedgerows, hedgerow
trees, thousands of acres of down and heathland, a third of our
woods and hundreds upon hundreds of ponds, streams, marshes and
flower rich meadows which have been systematically eliminated by
farmers seeking profit. In this the farmer has been aided and

abetted by Government, partly through price support, partly as a result of 'improvement' grants of various sorts.

Body is also concerned with this issue, though comments on land use feature only minimally in his writings. What he does discuss, however, goes much further by contending that there is also a diversion of capital to agriculture, which is actually causing unemployment elsewhere. In a memorably imprecise phrase he claims that 'The consequent loss of jobs by this diversion of £3000 million is incalculable: one million, two million, or is it more?' (Body 1982:125). Despite Body's claims to have the interests of farmers at heart, it is evident that when such statements are allied to the more explicit views of Shoard we are justified in speaking of British agriculture as under attack from a groundswell of opinion which is calculated to criticise not only governments but also farmers' organisations and farmers themselves. Indeed, farmers seem increasingly to be seen as enemies of the general interest.

My aim in thus characterising the current situation is not to argue the agricultural case. However, in seeking a measure of balance it may well appear that my remarks effectively constitute a critique of much of what has recently been written. Certainly, I wish to argue that before we become too ready to attack farming privileges it is important to look more closely at the realities of the position within agriculture, focussing particularly on income prospects and on the size of the sector. The story turns out to be a rather surprising one and it is only in the light of it that Body's main assertion can be adequately considered. The opportunity will then also arise to consider some of the views of Bowers and Cheshire whose book Agriculture, the Countryside and Land Use, while also being polemical, is more obviously rooted in economic analysis. I then turn to consider the position of the consumer and conclude with a brief look at environmental issues.

THE POSITION OF THE FARMER

The first point to be made in considering the position of the farmer is that, in the period since 1970, production and income trends have behaved in a rather complicated way. The indices used in this connection are somewhat tricky. We can begin (Table 1) with the concept of gross output which relates to all sales of products off the farm, inclusive of any (notably farm crops) later bought back as feed.

Table 1 Economic indicators for agriculture 1980 = 100

	Gross output at constant prices	Net product at constant prices	Net product RPI base	Farming Income RPI base	Productivity gross product per person
	(a)	(b)	(c)	(d)	(e)
1970	89	70	119	201	65
1971	92	78	122	207	72
1972	93	77	123	206	74
1973	93	78	148	263	75
1974	92	82	127	191	80
1975	89	72	125	193	75
1976	86	62	131	213	68
1977	92	77	118	181	79
1978	96	88	116	165	88
1979	96	87	110	132	88
1980	100	100	100	100	100
1981	100	104	103	116	105
1982	107	115	113	146	115
1983	106	108	104	119	112

a) Annual Review of Agriculture 1984 Table 24.
b) Annual Abstract of Statistics 1985 and earlier years.
c) Annual Review of Agriculture 1984 Table 23 b.
d) Annual Review of Agriculture 1984 Table 23 b.
e) Annual Review of Agriculture 1984 Table 24.

At constant prices (to give an indicator of volume) the index rose by 19 per cent from 1970 to 1983. Averaged out as a compound rate of growth it is 1.5 per cent per annum. This aggregate conceals a number of divergent trends though. The increase was higher for farm crops than for livestock products and static for horticulture, but

overall it is rather smaller than commonly supposed. If one goes
further to look at net product, which is essentially the national
income concept of value added, the position is somewhat different.
This deducts all inputs, including depreciation, from gross output
and is available in two forms.

The index which regularly appears in the Annual Abstract of
Statistics is intended to record the volume change in value added
using the familiar 'constant price' base for comparison of outputs
and non-factor inputs. Its movement over time (Table 1, column b) is
rather more rapid (there is a 54 per cent change from 1970 to 1983)
than that of gross output which is largely due to a change in the
efficiency with which non-factor inputs have been used. However, in
the Annual Review (Table 1, column c) it is presented in an
alternative form. Value added (i.e. net product) is calculated for
each year first at current prices and then adjusted by the retail
price index which is taken as an indicator of the 'purchasing power'
of agricultural factor incomes. The changes are dramatic. Peak
incomes were achieved in 1973 (48 per cent above the 1980 value),
while there has been a decline of 12.6 per cent over the whole
period. Using the same technique an indicator is also provided
(column d) for 'farming income'. This relates to the incomes for
their labour received by farmers and their spouses and from their
own capital invested after providing for depreciation but excluding
returns for land ownership. Though notoriously difficult to
interpret it provides a record of what are normally thought of as
farmers' incomes. Even allowing for the many problems and detailed
nuances of interpretation it is clear that the income base has been
markedly eroded in recent years. Though 1983 was a poor year real
incomes are now vastly below those at the start of the period, and
even more sharply down on the 1973 peak.

This may appear surprising in view of agriculture's well known
capacity to increase its productivity, and especially its labour
productivity. An index is provided in Table 2 column e in the form
published in the Annual Review. It is simply gross output (column a)
minus inputs excluding depreciation (hence it is in 'gross' form,
rather than the 'net' form of column b), divided by all labour
input. Given that the 'worker' input at 325,000 in 1983 (down 22 per

Table 2 Comparative price index numbers

	Average 1972-84	1980	1981	1982	1983	1984
Producer prices						
all	100	229	253	274	286	288
cereals	100	211	233	251	279	257
animals for slaughter	100	233	256	279	283	292
milk	100	242	266	291	295	292
Input prices						
all inputs	100	275	302	325	346	361
energy	100	373	450	513	572	570
fertilisers	100	287	317	332	335	344
feed	100	225	243	256	275	285
investment goods	100	322	348	375	395	411
labour	100	329	363	396	426	453
Terms of trade	100	83	84	84	83	80
Retail prices	100	275	307	334	348	364
Food prices	100	258	279	302	311	328

Source: Annual Review of Agriculture 1985

cent from the early 1970s) and the 'farmer' input at 203,000 (down
8.5 per cent) have both fallen it is not surprising that
productivity has been rising at a compound rate of no less than
4.8 per cent per annum. In effect this is the technical 'triumph'
recognised in Body's title. However, it has been accompanied by the
significant alteration in agriculture's 'terms of trade' also shown
in Table 2. Generally speaking the cost of inputs to agriculture,
except for feedstuffs which are largely constituted of agricultural
products, are determined by developments within the economy at
large. From Table 2 it can be seen that the index of all input
prices (up to 361 by 1983 compared with 1972-74) is close to the
change in the retail price index (up to 364). This, though, is
coincidental: as would be expected it is not true for energy,
markedly not the case for labour, nor indeed for investment goods.

The parallel, in fact, is maintained only by the behaviour of the price of feedstuffs. On the other side producer prices for all products have risen to an index value of only 288, and the 'terms of trade' (producer prices divided by input prices) have progressively declined to 80 per cent of their values in 1972-74. In relative terms farmers are receiving less for their output than they are paying for their inputs. Of course it might be suggested in relation to Table 1 that measures of income (e.g. column c), as opposed to volume of net product (e.g. column b) can differ: but they do so precisely because the income measure, when it is itself expressed in 'real terms', catches the terms of trade effect recorded in Table 2.

It should also be noted that Table 2 takes as its base the period when producer prices were particularly high; and that it straddles 1973 which was the peak year for farm income. What has happened, quite simply, is that the terms of trade effect has eroded the productivity gain. The reasons for the swing in the terms of trade are themselves complex and need not be considered in detail. Basically what has happened is that the exceptionally high prices of farm products in the 1970s have fallen back, largely under world market influences, and CAP determined prices - though above 'world' levels - have not been adjusted upwards in real terms. Indeed, the howls of protest which are always heard from outside the farm sector notwithstanding, the CAP has not in fact insulated farmers from a 'cost price squeeze'. There was, it is true, a bonanza in the early 1970s, but that period seems long since passed. Current indications, moreover, are that CAP price determinations (allied to limits on eligibility for support, which are the main aim of milk quotas for instance) will become increasingly less favourable to farmers.

This interpretation can obviously be disputed; but it is important to note it. Agricultural incomes can be subject to enormous swings and it is never easy to discover periods in which 'the trend will bend'. Indeed, as a cautionary tale it is worth pointing out that Bowers and Cheshire may well have fallen into just such a trap. Those who open their book will notice a chapter titled 'From rags to riches or what the government has done for farmers and what farmers have done for us'. It is a somewhat emotive heading to find in an academic work, but it does refer graphically to an

apparent switch in agriculture's fortunes in relation to the
relative income position in pre-war years which seemingly cannot be
disputed. Unfortunately (and perhaps rather surprisingly for a book
published in 1983) their main analysis ends with averages for
1973-77. In short, it catches the peak periods demonstrated in
Table 1. Moreover, as others have done, Bowers and Cheshire make a
great deal of the point that owner occupiers have also experienced
large gains from the rise in property values (Table 4.2 p87). But
again they end their analysis in 1973-77. There was some subsequent
increase to 1980 but since then the real value of farmland has
dropped by 15 per cent and with a declining real income base
property values seem destined to behave in a less spectacular way in
future.

However it can also be argued that although the agricultural
income base is far from buoyant in terms of its rate of change, this
gives no indication of income levels. This subject too is one which
is fraught with significant difficulties. Official figures are
available, but they are classified by type and size of farm using,
for the latter, a size unit which is unrelated to area. The figures
moreover usually refer to 'net farm income' which by definition
makes no deduction for imputed returns on working capital employed
in the business. The issue is further complicated by the fact that
many farm households have off-farm earnings or receive incomes from
non-agricultural uses of the farm house and land. There is also the
difficulty (mentioned above) in treating the position of long
established owner occupiers who may have substantial wealth
resulting from the increase in the value of real property.

Table 3, which shows the position for England in 1982/3, should
therefore be read with circumspection, particularly as there are
often large annual variations. For comparison, the earnings of UK
male manual workers averaged £7155 in 1982. The table also shows the
percentages of farms falling into each size-type group. If comments
can be hazarded it is clear that the smaller farmers, who make up
43.8 per cent of the total, are the least advantaged group while the
opposite extreme is represented by large dairy farmers (a small
percentage of total numbers) and large cropping farms which are much
more significant overall. Such points are well known and do not

Table 3 Net farm incomes, farm types and agricultural structure England 1982/3

Type of farm	Small		Medium		Large	
	£ per farm	Per cent total	£ per farm	Per cent total	£ per farm	Per cent total
dairy	6771	12.9	13734	14.9	29891	4.3
LFA cattle and sheep	7182	3.7	13075	1.3	n.a.	0.1
lowland cattle and sheep	1153	11.2	7142	3.0	18817	0.6
cropping	2446	12.4	10288	13.6	35424	13.0
pigs and poultry	n.a.	3.6	6831	3.0	30072	2.4
		43.8		24.8		20.4

Sources: Incomes from Annual Review 1984, with size categories defined in
European Size Units. The second column in each size group shows the
percentages of the total numbers of farms in the size group based on
June returns.

require extensive debate. What does emerge is the plain fact that
many farmers do not command the resource base to earn very high
incomes, and it is this which needs to be recognised in any
discussion of policy. It is a point which is particularly relevant
to discussion of Body's proposals.

RICHARD BODY - A BASIC SUMMARY

Richard Body's work has achieved the distinction of being the
subject of a prestigious conference organised by the Centre for
Agricultural Strategy and the Centre for European Agricultural
Studies (CAS and CEAS 1983). Two reasons seem to account for its
appeal. The first is Body's sheer skill as a pamphleteer; he weaves
a spell which is compelling despite the fact that the structure of
the argument is not always clear. Secondly, his proposals appear to
involve minimal damage for any particular interest group, and
potential benefit to some. Among the latter he has skilfully
interwoven concern for the Third World (and since what follows may
appear carpingly critical it must be conceded that much of what is
said on this difficult subject is sensible), with the interests of
major agricultural suppliers, and even pushes his case so far as to
suggest that British agriculture would gain from his proposals. The
general tenor of the argument is, perhaps, best characterised
presenting a series of direct quotations.

> 1 This book...chronicles a system of agricultural support that
> is expensive, uneconomical and unjust. It is a system that, for
> the sake of short term and often illusory gains is doing untold
> damage around the world. It cannot be abolished too quickly.
>
> 2 The evidence points to one conclusion. If the state had not
> interfered by restricting and taxing cereals from abroad, and
> if it had not spent taxpayers money on a policy that has had
> the effect of inducing farmers to grow arable crops rather than
> produce livestock, there would today be more and smaller farms,
> and more farmers and farm workers.
>
> 3 By re-opening our doors to the food which they (Third World
> countries) and the rest of the world are capable of producing
> more cheaply than we can, British agriculture would gain in the
> long term.
>
> 4 Arable farming would continue in a prosperous state on land
> suitable for it. Livestock farming would gain immeasurably.
> More farms, smaller farms and part-time farms would follow;

opportunities for the young farmer (now dismal) would be considerable, especially if he could afford to begin on his own in a small way.

5   This transformation in the farming pattern would make the countryside look different - more as we used to know it. Life in the countryside would be reinvigorated by the infusion of more farmers.

6   A system of grants and subsidies serves no purpose at all unless it distorts the price mechanism; and the only way to ensure that the naturally efficient producer receives the highest return is to allow the price mechanism to equate supply and demand.

7   The farmers in the main have not benefitted, strange though it may seem for many thousands of them have had to abandon farming, most of them with the utmost regret and reluctance.

If these statements are looked at carefully they reveal a thought process which can hardly be described as consistent. What Body is advocating is a rapid change in the support system (it cannot be abolished too quickly), which would have the effect of maintaining prosperity among arable farmers, but would actually increase the number of farmers in total, particularly in the livestock sector. We are further told (point 6) that the highest returns come from equating supply and demand, and that there would be a long term gain (point 3). This is peculiar indeed, if one wishes to maintain farming numbers - and this is all too obviously one of Body's aims - since any economist would predict that the ending of protection would have exactly the opposite effect. Surely it is very odd to make such an assertion for if farmers have not gained from support why should we suppose that they would gain from its abolition? The price mechanism, in the absence of protection, would not achieve the 'highest return' - far from it. The reader may feel immediately that this amounts to an unfair attack on Body, though, because what he is clearly envisaging is a transformation brought about in the livestock sector by reduction in feed prices.

Three important points need to be emphasised. In the first place it is clear from the price index material that feed prices are not out of line in trend terms with producer prices for milk or for animals for slaughter. This is never touched upon by Body and, in itself, suggests that he is overstating his case. Smaller farmers,

especially, are leaving the land. Secondly, since he is actively
canvassing the merits of reduced feed prices it might be thought
that he has a stronger point. After all, feed now accounts for some
34 per cent of the total costs of the average dairy farm so a 15 per
cent reduction in input cost on this item would bring an overall
fall of 5 per cent in aggregate cost. This would raise income, in
itself, by some 27 per cent, which is an apparently enormous change.
But we must remember that end prices for livestock products are also
supported (e.g. by levies and intervention purchases) and therefore
there is absolutely no guarantee that receipts would hold up if all
protection were abolished. To know what the total effect would then
be is a matter of conjecture but what makes Body's analysis worse is
that the point does not appear to have struck him. There is a third
telling point. If the net effect were to be favourable it does not
by any means follow that there would be an influx into farming. For
this to occur there would have to be division of holdings, for which
there would be little incentive.

Hence, for all these reasons, one must conclude that Body's
analysis is simplistic. Could we be certain (point 3) that British
agriculture would gain in the long term? It might adapt, and hence
become more economically efficient, but to pretend that the
'countryside would be invigorated by the infusion of more farmers'
is misleading. Furthermore, it is little short of scandalous to
pretend that a policy adaptation would benefit farmers as a group:
there would be survivors but that is all. My main conclusion
relating to Body's views, therefore, is that they are not only
unsound but dangerous as well. His policy would be unlikely to
produce the results which he envisages and a climate of opinion
which takes his scenario seriously should not be allowed to develop.

He has similarly distorted views on capital diversion as a
cause of unemployment in the economy at large. As we have seen he
argues that the loss of jobs elsewhere in the economy as a result of
diversion is incalculable using the phrase 'one million, two million
or is it more'? To blame agricultural protection for the current
slump is, frankly, unbelievable nonsense of the grossest kind. The
chapter on capital diversion in the earlier book and elsewhere is
good knock-about stuff: but it fails to distinguish between capital

formation proper, intermediate input costs, payments to non-land factors of production, and increased land values. Indeed, so vague is his accounting that he concludes with the lame statement that diversion of 'capital' must come to many thousands of millions of pounds, settling uncomfortably for around £3000 m. What he signally fails to demonstrate is the linkage between activities affecting a sector accounting for little more than 2 per cent of gross national production and an unemployment rate which has reached 13 per cent of the labour force. Nowhere does he mention adoption of 'monetarist' policies, cuts in public expenditure or the rise in the value of the pound in the early 1980s. His explanation is monocausal and completely blinkered. As so often happens in the book, however, he does hit on an issue which is important, only to distort its economic impact. One of Britain's failures has been a persistent tendency to misallocate resources between sectors and to support those which do not achieve significant impetus as engines of new growth.

A modest case can, in fact, be made against agricultural over-capitalisation though I would stress that it is modest. Real capital formation in agriculture, for example, defined as fixed investment (land values are excluded) has generally represented a somewhat larger proportion of total capital formation than agriculture's contribution to gross national product. But in very crude terms the 'excess' amount has been about 0.32 per cent of total capital formation, or about £126 m. per year at 1981 values, which represents only a minute percentage of GNP (about 0.07 per cent). Taken in conjunction with a capital-output ratio of 3, use in other industries would raise growth by 0.023 per cent per year. This is minimal, and merely serves to emphasise the fact that what may be a large misallocation in relation to a small sector may have trivial overall effects. It is certainly not the stuff of which major industrial collapse is born. One can only regret that Body's polemical skill has not been buttressed by a few shreds of economic literacy.

THE POSITION OF THE CONSUMER

At the start of this paper a contrast was drawn between the position of the consumer under the deficiency payments system of agricultural support and the state of affairs which now exists. It would be wrong to overestimate the extent of the problem however. At retail level about half of the price of foodstuffs is represented by processing and distributive margins which, in principle, should not be greatly affected by the source of foodstuffs entering the distributive process. Reference back to Table 2 shows that producer prices, as already mentioned, have been rising less quickly than retail prices overall. Food prices, at retail, have advanced by rather more than producer prices, but their increase is less than the change in the total for retail prices. In fact on a 1972-74 base the respective index figures are 328 and 364, so that the change for food is about 90 per cent of the total increase. I have detailed the behaviour of the long term trend in retail prices relative to other items entering the consumer budget since 1900 elsewhere (Peters 1980) so will not cover the issue in great depth here. All that is needed is a comparison between a relative food price index, and an index of prices of non-food items with the calculations set out in 'terms of trade' style. Although there are great difficulties in comparisons of this sort the trend is clear enough. At the turn of the century food was expensive in comparison with other goods then available, with the index standing at 150 (1980 = 100). Prior to 1914 Britain experienced a period of near stability in overall prices. However, while the country was geared to the pursuit of cheap food agricultural technology had not developed beyond a somewhat primitive and labour intensive stage and the most dramatic reductions in production cost had yet to occur.

During the first World War there was sudden inflation with food prices rising by almost 30 per cent in 1914 and continuing to escalate over the war years. Other prices also rose, though rather less rapidly, pushing the relative price index to 187 in 1916. Between 1913 and 1920 the disruption in major trade patterns, allied to the sharp post war boom, raised absolute food prices by about 2½ times. There was then a sharp reversal in trend and the boom ended reducing all prices, particularly those of food. By 1922 the relative index was down to 134, which was below pre-war levels. Minor alterations occurred in the remainder of the 1920s prior to

the next upheaval during the great depression of the 1930s.
Predictably food prices fell more rapidly than those of other
commodities and by 1932 the relative price reached its lowest
pre-1938 point of 106. There followed three short phases comprising
recovery to 1935, stability and a modest increase near the outbreak
of war. Post-1945 relative food prices fell sharply under the impact
of rationing, price control and subsidisation (the last at a cost of
£250 million in 1945, or about £1500 million at today's prices). The
ultra-cheap food policy continued in the three immediate post-war
years (the all time minimum for the relative price index was 89 in
1946 and 1948) and did not begin to crack until the 1949 devaluation
and the Korean War of 1950-51.

Somewhat later the phasing out of consumer subsidies and of
purchase on official bulk contracts, which had been important prior
to 1950, also raised prices. The end result was an absolute rise in
food prices of 50 per cent from 1949 to 1955 and a climb in relative
prices to 114 in the latter year. A downward trend then ensued.
Domestically, as we have seen, agricultural support was mainly based
on deficiency payments, and the period was also one in which the
developed world was in food surplus from which the British consumer
was able to benefit. In the later 1950s and the 1960s the United
States acquired large stocks of grain as a consequence of its farm
support policies, productivity in agriculture was advancing and by
the later 1960s the CAP was beginning to have an effect on world
markets, notably for cereals, as export subsidies were applied to
enable European producers to supply third countries.

Despite the 1967 devaluation Britain's food in 1970 was cheaper
in relative terms (index value 103) than in the depths of the 1930s
depression, though it was not down to the levels associated with
heavily managed markets in the late 1940s. There were then some
increases, but these were as much associated with the 'world food
crisis' of the early 1970s as with accession to the EEC. The peak
was reached by about 1978 (index value 106.3), but the downward
trend was subsequently re-established to reach 93.0 in 1982. There
remains considerable controversy about the effects of the EEC on
retail food costs. Official Ministry of Agriculture figures placed
retail prices about 8-10 per cent higher than they might have been

if Britain could have bought its import requirements at what was
then the current level of 'world' market prices, and there are
numerous similar estimates. While it is important not to minimise
these effects it is nevertheless crucial to realise that the long
term trend remains. Neither do further developments in the CAP, with
its prudent pricing policy, suggest that rising relative food prices
will become a major issue.

THE COUNTRYSIDE ISSUE

To conclude this selective review of the attack on agriculture only
a few comments need to be added to the spate of books, papers and
conferences which have recently addressed the issue of the
'countryside'. Many will feel that this is a matter on which the
critics, including Body, are on firmer ground and there is much to
agree with in the main thrust of the arguments which have been put
forward both by conservationists, whose aim is preservation, and by
others more concerned with access and recreation. Economists are in
an unhappy position in this debate; not because they have nothing to
say in principle, but because the 'values' involved are difficult to
define, and even harder to quantify. There is a vast range of
economic theory relating to 'externality' problems, optimal taxation
schemes to mitigate their effects can be proposed (e.g. specific
taxes on nitrate inputs) and there are tools of cost-benefit
analysis which might be applied as proposed, among others, by Bowers
and Cheshire.

Despite their having fallen into the trap outlined above there
is much to admire in Bowers and Cheshire; they pull a mass of well
known material into a cogent argument; rightly pour scorn on the
idea that a cut in support would intensify production; and argue
against the 'planning controls' advocated by Shoard favouring,
instead, a switch in economic incentives which would be
self-policing and thus less bureaucratic. To quibble a second time
however their cost-benefit work on conservation is not flawless.
Basically they consider situations in which land is to be
'reclaimed' for agriculture (e.g. by the drainage of marshland) and
seek to show that the value of the resultant output (which
essentially measures the opportunity cost of 'conservation') should

be measured at 'world prices' rather than at support values. Since
Britain is not 'self sufficient' in indigenous foods (the figure is
now about 80 per cent) increased output would, very broadly,
displace imports. If these could be obtained at 'world prices' their
argument would be correct. Indeed, I have used it myself in a
different context (Moore and Peters, 1965).However, any imports can
now only be obtained at CAP price levels; either by direct purchase
from Europe, or by contributing levies on imports to the CAP budget.
The argument can, needless to say, become very detailed: suffice it
to say now that opportunity cost cannot easily be equated with the
'world price'.

It is also very wrong to regard economists in general as being
interested solely in production as such for the god that is
worshipped is not so much the size of gross national product, as
optimality in resource allocation. And in relation to that more
important question there are hopeful signs that administrative
machinery is already in place to deal with countryside issues
principally in the form of the National Parks, the Nature
Conservancy Council, the Countryside Commission and the recently
tightened provisions of the Wildlife and Countryside Act 1981. The
National Farmers Union has, meanwhile, expressed sympathy with the
general tenor of the conservationist argument in its policy
statement New Directions in Agricultural Policy: The Way Forward
published in September 1984. My real problem in looking at the
raging debate is a feeling that while there are evidently many
specific issues which need to be dealt with much of it centres
around that inherently indefinable quality - the 'appearance' of the
countryside. To suggest, as Body does, in one of the quotations
presented above that his policy (essentially of less intensity on
the arable side) would make the countryside 'more as we used to
know it' could be the merest pipe dream. We are inclined to forget
too easily that agricultural prosperity can aid 'appearance' and
that lack of it can be destructive. When faced with cries of anguish
it is useful to remember the words of C.S. Orwin in Speed The
Plough. Written in 1942 the book graphically describes agricultural
conditions in the inter-war period - the low earnings of farm
workers and farmers, the burden of taxation and so forth. In support

of all this Orwin maintained:

> we have the evidence of our eyes, for it is impossible, as we
> move about the countryside, to overlook the bad repair of so
> many of the farm buildings, the absence, almost complete, of
> new buildings, the mire and muck in which the farmer and his
> men are obliged to work; the hedges overgrown, the ditches
> silted up, the gates broken on so many holdings; the
> blacksmiths' and the wheelwrights' shops closed in the
> villages.

Contrast this with Body who believes that it is a myth that
British agriculture was in a state of depression by the time the
1939-45 War began. He makes some concession to depression in arable
farming but states that 'the livestock sector flourished - and
flourished as much as any branch of our nation's trade and
industry'. All our views on 'appearance' must be subjective to some
degree; but Body cannot be allowed to make emotive assertions which
run so contrary to the observations of Orwin, one of the most
distinguished inter-war commentators. It is a moot point to debate
the merits of the current scene against that of pre-war Britain.
What is important is that we do not forget that the 'traditional'
scene, after which so many appear to hanker, had its drawbacks.
Looking at the past through rose coloured spectacles can all too
easily lead us to undervalue the achievements of the present.

CONCLUSION

At the outset I stated that the aim of this paper was to do no more
than attempt to put some of the current controversies into a more
balanced context. The argument I have presented can be summarised in
the following way. A priority when examining 'the state of
agriculture' is to remember that the income position of farmers is
now under pressure, and this is particularly so among smaller
producers. The current squeeze is a very real one and though
sympathy for the 'dairy dukes' and 'barley barons' might be limited
it is as well to remember also that there are relatively few of
them.

Attacks on 'farmers' are only too easy to mount, but we can all
find economically active groups within the population who appear to
enjoy a position of unwarranted protection. The wildest critiques,

exemplified by the work of Body, may have elements of justification within them; but it can hardly seriously be pretended that a sharp and immediate cut in support would benefit <u>all</u> farmers and the economy at large. There is, to be sure, a case for re-aligning support mechanisms and the work of Black and Bowers (1981) can be cited in support of such ideas. A similarly cogent argument, with some sympathy for those facing severe adjustment, is provided by Howarth (1985). By comparison the food prices issue must be considered relatively minor once it is considered in the light of long period trends. Carefully presented there is a 'conservationist' case to be answered but too often it is stated in a highly polemical and subjective fashion. But whatever the merits of specific elements of the case I have been concerned to show that the general assumption that a return to lower protection would be immediately beneficial is likely to be naively mistaken.

REFERENCES

Black, C. J. and Bowers, J. K. 1981. The level of protection of U.K. agriculture. University of Leeds: School of Economic Studies Discussion Paper No 99.

Body, R. 1982. Agriculture: the triumph and the shame. London: Maurice Temple Smith.

Body, R. 1983. Farming in the clouds. London: Maurice Temple Smith.

Bowers, J. K. and Cheshire, P. 1983. Agriculture, the countryside and land use. London: Methuen.

CAS and CEAS, 1983. Agriculture: the triumph and the shame, an independent assessment. University of Reading: Centre for Agricultural Strategy.

Howarth, R. 1985. Farming for farmers? London: Institute of Economic Affairs.

Moore, L. and Peters, G. H. 1965. Agriculture's balance of payments contention. Westminster Bank Review, August.

National Farmers' Union, 1984. The way forward - new directions for agricultural policy. London.

Orwin, C. S. 1942. Speed the plough. London: Penguin.

Peters, G. H. 1980. An outline history of British food prices. Oxford Agrarian Studies, 9.

Self, P. and Storing H. J. 1962. The state and the farmer. London: George Allen and Unwin.

Shoard, M. 1980. The theft of the countryside. London: Maurice Temple Smith.

11 Agriculture and conservation in Britain: a policy community under seige[1]

GRAHAM COX, PHILIP LOWE and MICHAEL WINTER

'The Crown is one and indivisible: at least in constitutional
theory.' (House of Commons Environment Committee, 1985.)

INTRODUCTION

Contemporary analysis of British government suggests that in most
sectors of industrial, economic and social policy there exists a
'policy community' in which key interest groups enjoy a more or less
close partnership with the relevant government departments and
statutory bodies in the formulation and implementation of policy. As
a result even Parliament itself may play little direct role in the
policy process (Jordan 1981, Jordan and Richardson 1982). Moreover
analysis suggests that, taken together, a number of factors have
meant that over the past decade or so most of these policy
communities have faced growing external pressures.

Efforts to both co-ordinate and integrate policy sectors and
recognize the interdependence of policy problems have, first,
challenged the isolation of policy communities. Heclo (1978), in
particular, has described how, as a consequence, the number of
participants in any particular policy area has tended to grow, with
the result that it may be difficult to discover the locus of power
in what he terms these 'issue-networks'. Such developments have,
therefore, reduced the extent to which policy communities occupy
distinct and relatively well insulated domains or 'political
spaces'. Secondly, policy communities have also faced pressure from

newly formed or newly politicised interest groups. Marsh (1983) has summarised the changing nature of British interest groups during the past two decades. He points to the multiplication of the number of interest groups, particularly those of a promotional and ideological or single-issue kind. At the same time already existing groups, notably economic ones, have become more politically involved.

In addition groups have increasingly emphasised contacts with the Executive and have consequently downgraded contacts with parties and Parliament. There has been a consequent increase in the formalisation of contacts between interest groups and government. New groups have, moreover, challenged the authority of policy and have tried to force their way into established policy communities. The proliferation of groups has led to considerable 'overcrowding' in some policy communities with decision makers finding it very difficult to co-ordinate and resolve the various pressures put upon them. Many of the newer citizen and public interest groups have also pursued unconventional means of participation, often demanding new participatory structures and new forms of policy analysis, which have placed considerable strains upon established procedures.

Some policy communities have been relatively receptive to these outside pressures which have been as evident in relation to the implementation of policy as its formulation, whilst others have staunchly resisted them. Outsider groups adopting unconventional tactics have shown that they can often effectively block hitherto established policies. Environmental groups, for example, have delayed or prevented the construction of roads, airports and power stations. Such groups, however, typically lack the power to impose their own alternative solutions. The consequence may be 'pluralist stagnation' and 'institutional paralysis' (Wootton 1978) with policy makers in various sectors prevented from implementing conventional policies and yet unwilling to grasp alternative ones. The re-running of public enquiries, for instance at Stansted Airport and London's Archway Road, seems to be one rather bizarre symptom of such paralysis. Equally bizarre was the stalemate, over the routing of the proposed Okehampton by-pass, consequent upon the 1985 decision taken by a special parliamentary committee contradicting the decision of the earlier public enquiry.

AGRICULTURAL AND CONSERVATION POLICY SECTORS AND COMMUNITIES

The Conservation Policy Community[2]

Although the policy community for rural conservation is especially
large and diverse it is possible to make a rough estimate of its
size. Michael Heseltine, when Secretary of State for the
Environment, ordered a review of the relationship between his
department and environmental groups, and as part of this exercise a
list was drawn up of pressure groups associated with the Department
of the Environment. It includes 130 national organisations concerned
with aspects of the 'natural environment'. A number of these are not
environmental groups but are organisations concerned with the impact
of environmental policies upon their members. These include the
Confederation of British Industry, the National Farmers' Union and
the Country Landowners' Association. Conversely the list, by
confining itself to national groups, does not fully reflect the
extent of environmental pressure as exerted by an additional large
number of local groups.

The two major quangos in this field, the Nature Conservancy
Council (NCC) and the Countryside Commission, also have their own
constituencies. For each an umbrella body exists to co-ordinate the
environmental groups lobbying the two agencies. Wildlife Link has a
constituent membership of 33 national and international
organisations. Its counterpart, Countryside Link, has a membership
of 13 national organisations.[3] In addition the Countryside
Commission holds an annual conference of voluntary organisations
with which it works, and this brings together representatives of
some 40 organisations. Of course the names of certain of the groups
recur on each list, and by the same token members and key
individuals may be involved in more than one group, but it is
nevertheless clear that the policy community is large, involving up
to 200 organisations. It is a community that has mushroomed in the
past 20 years as pressures on the countryside have grown leading to
intense competition and conflict. Gradually groups have organised
themselves to protect or promote their particular interest. The
rapid spread of the county trusts for nature conservation in the

1960s from 10 trusts at the beginning of the decade to 39 at the
end, exemplifies such developments.

Other interests have responded in like manner. The British
Field Sports Society, set up specifically to resist legislative
restrictions on for example fox hunting and shooting, has been
obliged to adopt a higher political profile in recent years.
Voluntary organisations, which once existed chiefly to service the
specialist interests of their members, have taken on a political and
representational role. Examples include the Caravan Club and the
British Association for Shooting and Conservation. Pressure groups
once quiescent, like the Council for the Protection of Rural England
(CPRE), the Ramblers' Association and the Council for National
Parks, are now active again, in the case of the CPRE adopting a
considerably more radical stance on a number of issues. New groups,
such as the British Hanggliders' Association and the Byeways and
Bridleways Trust, have emerged to represent novel concerns. In
addition some of the new-style public interest groups previously
pre-occupied with other concerns, such as Friends of the Earth, have
turned their attention to the British countryside. Finally the
conservation policy community contains a number of those rural
economic interest groups which have long been politically active in
other spheres.

The Agricultural Policy Community

In contrast to rural conservation the agricultural policy community
remains a remarkably closed one. It is a community in which a small
number of interest groups, pre-eminently the National Farmers' Union
(NFU) and the Country Landowners' Association (CLA), are highly
significant, in some instances enjoying exclusive representation
enshrined in statute. Thus the 1947 Agriculture Act required the
government, when setting agricultural price support levels, to
consult 'such bodies of persons who appear to them to represent the
interests of producers in the agricultural industry'. This has been
taken to mean the NFU and its Scottish and Ulster counterparts and,
since 1978, the break-away Farmers' Union of Wales. The CLA, which
represents the wider interests of owners of rural land, whether
owners of tenanted land or owner-occupiers, has through its careful

cultivation of contacts similarly enjoyed relatively easy access to
the state.

Given this statutory role in the annual farm price review
agricultural interests developed and consolidated their governmental
links, particularly with the Ministry of Agriculture, Fisheries and
Food (MAFF). Already by the early 1960s the relationship between the
NFU and government was, according to Self and Storing (1962),
'unique in its range and intensity'. Later commentators have tended
only to confirm and amplify this judgement. Indeed, in recent
analyses of British politics, agriculture has been cited as perhaps
the one unequivocal example of an economic sector where an interest
group has been officially recognised by the state and incorporated
into the process of decision making, not merely to represent its
members but to play a joint role in the political management of the
sector. (Grant 1983, Metcalfe and McQuillan 1979, Richardson and
Jordan 1979, Wilson 1977.)

The NFU has derived considerable political advantage from its
symbiotic relationship with MAFF: through the Ministry's
single-minded commitment to the farmers' cause, through the NFU's
entrenched role in policy making and through its privileged access
on a routine basis to centres of decision making, including the
highest levels of government. The Ministry and the Union are in
constant contact at all levels over the myriad of issues, large and
small, that arise in the development and implementation of policy.
The NFU's working partnership with the state has enabled it to exert
an influence which has been disproportionate, given that its
membership is too small to determine directly the outcome of
elections and it does not have at its disposal the direct economic
sanctions available to some other key business organisations or
trade unions.

During the past decade, however, various factors have ensured
greater prominence and contention for questions of agricultural
policy. These include the excesses of the Common Agricultural
Policy, the shift of the burden for agricultural support from the
taxpayer to the consumer which followed EEC entry, rising unease
over animal welfare and political concern over the social welfare

of rural communities. But the most serious challenge to the privilege of farmers arises from conservationists (Lowe et al 1986). As the Society for the Responsible Use of Resources in Agriculture and on the Land (RURAL) noted in the report of its inaugral meeting held in April 1983 'farmers are becoming increasingly aware that they are under threat to their freedom of action because of the strength of public opinion about the way land is being managed' (Wilkinson 1983).

The Autonomy of the Farmer and the Agricultural Policy Community

In recent years the corporate groupings which represent the interests of farmers and landowners and which arise from their position with respect to the division of labour in the sphere of production, have faced opposition from conservation groups, which, as associations of individuals with common interests, inhabit the very different competitive sphere of pluralist politics (Cox and Lowe 1984). The distinction between 'corporate bias' (Middlemas 1979) in the realm of class politics centred on production issues, and the non-class competitive struggle in the sphere of consumption has been highlighted by Cawson and Saunders (1983). Whilst the distinction is not without its problems it does represent an advance upon positions which argue that all political struggles must be reduced to their class determinants. And by utilising concepts such as 'corporatism' and 'pluralism' to describe historically specific arrangements co-existing in different sectors of the same polity we avoid repetition of the sterile theoretical wrangling between proponents of corporatist or pluralist theories of the state.

In responding to environmental criticisms of modern farming, agricultural interests have been determined to preserve two cherished and related freedoms: first the autonomy of the Ministry and of the farming community in the administration and implementation of agricultural policy; and second the autonomy of the farmer in making production and land use decisions. For environmental groups are not only seeking a voice in agricultural policy but also a more detailed legislative framework for land use management. Thus the interests of the farming and landowning community and MAFF coincide in resisting the encroachment of environmental groups and any ensuing environmentally motivated constraints on their freedom of action. Of these two areas of concern by far the most compelling has been the desire to sustain the integrity of the policy community.

This may seem a strange claim given the emphasis, particularly in NFU ideology, on the vital need to maintain the freedom and scope for initiative of the individual producer. What is significant, however, is the willingness of the NFU to sacrifice this principle and accept controls over farmers' actions when these are operated by the Ministry or its agents. Thus the NFU accepted the various controls, including supervision orders and powers of eviction, embodied in the 1947 Act, and unsuccessfully resisted their repeal in 1958. It has supported the introduction of stringent controls in a succession of disease eradication campaigns and has assisted in policing them. It has co-operated closely with commodity marketing boards in the imposition and implementation of a range of constraints on farmers in growing and selling their products.

The Union supported a compulsory levy on employing farmers to finance the Agricultural Training Board, something objected to by the Farmers' Union of Wales. It accepted the creation of the Northern Pennines Rural Development Board in 1969 notwithstanding its powers to intervene in the land market and in land use change. When the Board was wound up two years later the Union lobbied for its re-introduction. Similarly it has pressed for the re-introduction of the requirement that farmers should obtain the Ministry's prior approval before embarking on improvements likely to be eligible for agricultural capital grant, a requirement removed in 1980 following a Rayner review.[4] More recently the NFU, while objecting to the speed and manner in which it was done, has acquiesced and actively co-operated in the imposition of milk quotas, despite some voluble criticism from its own members and the condemnation of the Farmers' Union of Wales and the National Milk Quota Action Group set up to rally the opposition of disaffected producers.

This ready acceptance of a range of controls on farmers' actions suggests the need to qualify Grant's claim (1983) that a

major advantage for the farming community of the NFU's corporatist
association with the state is that 'legislative control of the
activities of farmers is used only as a last resort'. Having
advanced that claim Grant (1983) goes on to argue that 'a
corporatist arrangement can only work if the NFU is able to
discipline its own members so that they abide by agreements arrived
at with government and generally co-operate with the implementation
of government agricultural policy'. But what such an argument
overlooks is the fact that the NFU finds it expedient, in some
cases, to accept or even advocate the imposition of formal controls
on farmers, usually as part of an overall package which includes
benefits and incentives. What it gains thereby is considerable
influence over both the form of controls and the manner of their
administration. At the same time it avoids the strains on its own
authority that any attempt to discipline its own members would
entail. Indeed it is evident that the farming lobby is quite willing
to sacrifice elements of farmer autonomy to achieve a policy goal
which benefits the farming community and which maintains or extends
its own power. While this argument clearly applies within the arena
of agricultural production policy it does not extend to the
environmental field. Environmental regulations, by making farmers
answerable to non-agricultural authorities and interests, would have
the opposite effect of diminishing the power of the farming lobby.
Such regulations have therefore been tenaciously resisted. By what
means and to what extent MAFF and the farming interest groups have
managed to maintain the integrity and exclusivity of the
agricultural policy community in the face of mounting criticism from
environmental interests are thus critical questions.

REPERTOIRES OF RESISTANCE AND STRATEGIES FOR CHANGE

Defending the Agricultural Policy Community

First of all it must be emphasised just how closely the NFU, CLA and
MAFF co-operate in defending their common interests. Clientelism[5] is
the term used by Richardson and Jordan (1979) to describe the
relationship between a government department and its lobby when
there is a close identification of interest between the two. It is
perhaps particularly appropriate to use the term in relation to the

agricultural sector, where the relationship can be so close that it
often appears to be one of equivalence rather than hierarchical.
Clientelism, in short 'seems to mean that the departments articulate
the views of their clients but sometimes it means that the pressure
groups control the department. But really the latter interpretation
is only a hostile version of the former' (Richardson and Jordan
1979). Similarly the relationship between MAFF and the NFU has been
characterised as 'symbiotic' by Newby (1979). For Wilson (1977) 'the
MAFF is not the ruthless servant of the NFU, but its normal way of
doing business helps the Union's cause'. Richardson et al (1978),
meanwhile have described how MAFF is able to mobilise the support of
the NFU and the CLA in defending its administrative territory.

This is not an unusual stance for any department or bureaucracy
to take towards administrative changes, but defence of
administrative territory has special significance for MAFF and its
associated interests. Agriculture is the only industry to have its
own Ministry. This not only gives the industry crucial access to
Cabinet decision making and the allocation of resources, but it also
underpins the exclusive relationship between the Ministry and
agricultural interest groups, although the Ministry also has
responsibility for the food-processing industry whose interests are
often at odds with its principal supplier of raw materials.
Richardson et al (1978) have specifically examined the jostling
between the Department of the Environment (DoE) and MAFF over
responsibility for land drainage. They comment:

> There was considerable sensitivity within MAFF that if land
> drainage was lost, the rot would set in and other functions
> would be lost to other departments, for example...health and
> safety, and education and training.... In theory a number of
> MAFF's functions could quite logically be transferred to other
> departments, thereby undermining the rationale for MAFF's
> existence.
>   (Richardson et al, 1978:54).

Recently such proprietorial concerns have assumed much greater
salience with alarmist speculation in the farming press that the
current Minister of Agriculture may be the last (Big Farm Weekly,
17 May, 1984). However the priority attached to the defence of
established territory does not necessarily extend to the Ministry
desiring to expand its remit which might place undue strain upon its

own resources. A delicate balancing act is required by the Ministry in its insertion into the ebbs and flows of policy processing and implementation. Thus in the case of conservation MAFF and its client groups have assiduously opposed the imposition of constraints which might give outside groups and departments leverage over its work. Thus it failed to give any form or substance to the requirement of Section 2 of the 1968 Countryside Act which provides that, when exercising a function in relation to land, every Minister and government department must have regard to the natural beauty and amenity of the countryside.

The Ministry also successfully blocked attempts to make this duty more specific in the passage of the Wildlife and Countryside Act (Cox and Lowe 1983a, 1983b). More recently Dr. David Clarke's Private Members' Wildlife and Countryside (Amendment) Bill sought, inter alia, to place a duty on the Ministry of agriculture to further conservation. Though it received an unopposed second reading in the House of Commons in February 1985 Government amendments at committee stage removed the clause.[6] In the words of William Waldegrave, Under Secretary of State at the DoE, 'it would be a logical absurdity'. Elaborating upon this point in the correspondence column of the magazine Country Life he supplied a rationale for MAFF's evident reluctance to endorse a specific commitment relating to an activity with which it would claim to have been concerned at least since 1968:

> Why should not my Department have a duty to further
> agriculture, or the Foreign Office to further the treaty
> obligations on endangered species? There would be no end to
> it... Conservation is government policy, not MAFF policy or DoE
> policy.
>
> (Country Life, 11th April, 1985:956).

In fact the phrase 'logical absurdity' hardly seems warranted, even in the light of the formalities of constitutional theory. Section 41.3a of the Wildlife and Countryside Act already lays a specific duty on agriculture ministers 'to further the conservation of wildlife and the enhancement of the natural beauty of the countryside' in carrying out certain duties within designated Sites of Special Scientific Interest, National Nature Reserves and National Parks. The Clarke amendment sought to extend the duty to include the wider countryside. In terms of the analysis being developed in this paper it is perhaps most appropriate to see the Under Secretary's apparent defence of MAFF as indicating - whatever else - a concern to defend DoE territory. However the defence of narrow departmental lines of demarcation belies the confusion, duplication and gaps which already exist in the implementation of conservation policy at the local level.[7]

It is perhaps this confusion at the 'sharp end' which has prompted the agricultural interest groups, less sensitive to questions of departmental resources, to adopt a different line to the Ministry on this matter. The administrative reforms favoured by the groups would have the effect of expanding MAFF's administrative territory. The most radical proposal has come from the CLA, which has suggested a reorganisation of central government responsibilities to create a Ministry of Rural Affairs. The core of the new ministry would be MAFF to which would be transferred the DoE's responsibilities for rural planning and development, and oversight of the NCC, Countryside Commission and Development Commission. Conservation groups have rejected such a move fearing the subordination of environmental interests to agricultural interests. In particular they fear that conflicts currently aired on an inter-departmental basis would be internalised within one department and stifled by the dominant interest and the dead hand of official secrecy. The NFU have also opposed the idea but did support the Clarke amendment resisted by the Ministry.

The Ministry of Agriculture's Advisory Council for Agriculture and Horticulture in its 1978 Strutt Report Agriculture and the Countryside (MAFF 1978), proposed more modestly that the remit of MAFF's Agricultural Development and Advisory Service (ADAS) should be widened to include conservation as well as food production. In the event few of the Strutt Report's recommendations reached the statute book. However since before 1970 ADAS had been developing its conservation expertise, and in presenting evidence to a House of Lords Select Committee (1984a) early in 1984 Lord Belstead, Minister of State for Agriculture, detailed the development of the environmental expertise of ADAS. But notwithstanding the Ministry's spirited defence of its contribution to the improvement of the environment, the Committee took the view that both MAFF and the DoE (which it felt had in the past been largely subordinate to MAFF) had been insufficiently responsive to the large and reasonable body of opinion calling for action to bring the impact of rapid change in the countryside under control.

Subsequently the Countryside Commission, among others, has voiced similar concerns in a memorandum to the House of Commons Environment Committee (1985). The conservation interest, it suggested, had been slow to develop within ADAS and neither it, nor the provision of socio-economic advice, had been adequately integrated within the service. The Commission expressed the hope that the review of ADAS conducted by its new Director General Ronald Bell would propose a strengthening of ADAS's conservation role. The published report (Bell 1984) does, indeed, suggest a continuation of the trend which has seen ADAS gradually build up a modest degree of conservation expertise in its Land and Water Service (LAWS).[8] However some conservation organisations have expressed disappointment with the vague generality of its statements on conservation. The CPRE, for instance, regretted that Bell's statement of the principal function of ADAS was unqualified by any reference to the environment or the wider community and it remains the case that conservation courses for ADAS officers are aimed largely at LAWS officers who constitute only 13 per cent of ADAS staff. There is clearly a danger that LAWS, while becoming a specialist conservation service within ADAS, might reach only the already interested farmers. There seems to be no recognition in the report that if conservation is adequately to be integrated into the work of ADAS then the Agriculture Service too needs to have similar responsibilities towards conservation. The overall impression of there being only a minor adjustment of attitudes seems confirmed, moreover, by the absence of any reference to conservation interests or objectives in the report's discussion of ADAS's Research and Development programme.

There must be doubt, therefore, as to whether the recommendation of the House of Commons Environment Committee (1985) that 'conservation be given a greatly increased priority in the training and work of ADAS staff' will be taken up to the extent required. In the face of broadly based public feeling, expressed over many years, MAFF has, by responding with a combination of apparent disregard and studied gradualism, consistently worked to affirm the existing order, defending meanwhile its distinctive administrative territory. It has to be remembered, of course, that some provision for conservation was made at the end of the 1960s. This was as a response to the provisions of the 1968 Countryside Act and the personal goadings of the then Minister of Agriculture, Cledwyn Hughes. What is surprising is how slowly things have changed since then, in spite of the transformation in public concern on such issues.

The Contrasting Tactics of the Agricultural Lobby and the Ministry

As well as putting over their own proposals for administrative and policy reforms, the agricultural lobby has, of course, sought to influence the terms and nature of the conservation debate itself. There is a clear distinction which can be drawn here between the NFU and CLA on the one hand and the Ministry on the other. The interest groups have been more outward going, and being challenged, perhaps more than any other economic interest group, by environmental criticism, have actively engaged in discussion with conservation groups. Throughout they have stressed the need to foster and maintain the voluntary co-operation and goodwill of the farming community if practical remedies are to be found in response to conservation problems. Many farmers and landowners are indeed keenly aware of the claims of conservation and the CLA has long reflected this interest.

The NFU, by contrast, formerly showed little interest in matters which, though a suitable avocation for leisured landowners, seemed marginal to the concerns of the working farmer. However since the first attempts to introduce some statutory regulation of agricultural development in environmentally sensitive areas in the 1968 Act, the NFU has joined forces with the CLA in emphasising its concern for conservation. By the same token it has ensured representation of the farming viewpoint in conservation agencies (Brotherton and Lowe 1984). Together the NFU and CLA have responded to the charges of conservationists by presenting farmers and

landowners as stewards of the countryside (Cox, Lowe and Winter
1985).

In practical terms this has meant considerable effort by the
two organisations to refurbish and broadcast a stewardship ethic
among their members. As the foreword to a 1984 handbook on
management agreements issued jointly by the NFU, CLA and Royal
Institute of Chartered Surveyors takes care to emphasise, landowners
and farmers have moral responsibility for the sensible stewardship
of their property. The CLA, in particular, has worked to foster this
sense of moral obligation amongst its members giving conservation
issues considerable prominence in its monthly journal The Country
Landowner. But though it has ensured an exceptionally high profile
for pressing environmental matters within its constituency the
farming and landowning lobby has, when confronted by specific
environmental problems, typically responded by promulgating schemes
of voluntary self-regulation. It has, moreover, advanced claims for
the effectiveness of such schemes when countering demands for more
formal controls over farming activities. Agriculture's exceptional
freedom from many statutory environmental regulations and the
reliance placed instead upon informal or voluntary safeguards is an
indication of the skill with which agricultural autonomy has been
defended and the significance of the post-war policy commitment to
increased agricultural production.

However resources for the defence have been harder to mobilise
of late, nowhere more so than in the defence of agricultural
autonomy during the passage of the Wildlife and Countryside Act (Cox
and Lowe 1983b). Having ensured that its commitment to voluntary
co-operation was enshrined as the fundamental principle of the Act
the farming and landowning lobby has faced the task of convincing a
sceptical conservation lobby and ensuring that the practices of the
industry live up to the rhetoric of the goodwill case (Cox and Lowe
1983a, Cox, Lowe and Winter 1985). But while there is considerable
public attention focused upon issues relating to nature
conservation, agricultural exceptionalism is, in fact, evident
across the whole range of environmental regulation. In some cases
farming activities have been specifically exempted from control as
in the case of statutory land-use planning. In other areas, special

voluntary arrangements or procedures quite distinct from the
statutory regulations imposed on other industries have evolved, as
for example in the Pesticides Safety Precautions Scheme. Indeed the
place of formal controls has tended to be taken by codes of conduct
such as the NFU's Codes of Practice on Straw Burning, Aerial
Spraying and Silage Effluent.

In some cases codes of conduct have resulted from the
implementation of legislation, producing the paradox of a statutory
yet voluntary code. Examples include the Code of Guidance for Sites
of Special Scientific Interest issued under the Wildlife and
Countryside Act and the Codes of Good Agricultural Practice to be
issued under Section 31 of the Control of Pollution Act 1974. The
Ministry supplements such codes by making freely available an
extensive range of advisory leaflets and booklets on issues of
concern such as building design, farm waste management, the use of
fungicides and insecticides and all aspects of pollution control. In
addition, in a number of regions, it provides a telephone
information service. Considerable resources, therefore, are devoted
to enabling farmers and landowners to operate in a responsible
manner.

The NFU and CLA have similarly put a good deal of effort into
promoting these codes with their members. Exhortation to abide by
them is generally reinforced by warnings of what might happen if
voluntary procedures patently fail. The CLA, for example, has
produced its own Code of Guidance on SSSIs and urged its members to
comply 'in both the letter and the spirit'. Its magazine warned that
'if performance of the landowner is disappointing, it could do more
than anything to shake confidence in private ownership' (Country
Landowner, September, 1981). The NFU likewise has circularised its
members warning them that:

> the Government, with NFU support, has a very tough task in
> maintaining this largely voluntary code as opposed to the
> range of compulsory restrictions proposed by conservation
> groups. It must be stressed that if farmers are not prepared to
> be conciliatory on SSSIs, for example by modifying schemes
> where possible and seeking management agreements (with
> payments) where appropriate, then there can be no question that

a future Government of any party would consider more punitive
controls for SSSIs and possibly elsewhere.

(NFU Press Release, 6 November, 1981).

While the NFU and CLA use the fear of external controls to
press their members into line, they invoke a different spectre to
convince environmentalists of the appropriateness of voluntary as
opposed to statutory safeguards. Implicit, and often explicit, in
much of the NFU's and CLA's argument against controls is the claim
that these would be unworkable because farmers and landowners would
not co-operate in their implementation. They even paint the rather
unlikely picture of agrarian lawlessness and recalcitrance, with
farmers taking disruptive or pre-emptive action. The President of
the NFU, Sir Richard Butler, for example, has warned
conservationists that to press for controls would be
counter-productive, claiming that already natural features of the
countryside have been deliberately destroyed by farmers alarmed at
the possibility of planning constraints. 'Push farmers into a
corner with nowhere else to go and they will go for your throat' he
cautioned. 'Rest assured that a significant proportion will get
their tractors out and rip up perhaps a grove of orchids, meadowland
and hedgerows out of sheer frustration and annoyance' (Daily
Telegraph, 26 March, 1984 and Farming News, 30 March, 1984).

The missionary work of its client groups has allowed MAFF to
adopt an apparently impassive stance. This is in contrast to other
government departments, which through the 1970s adopted new
consultative and evaluatory procedures such as the establishment of
advisory committees and commissions including environmental
representatives, in response to environmental pressures. So far MAFF
has staunchly resisted any move to open up the agricultural policy
community in this way by giving consultative status to environmental
interests. The Bell Report (1984), for example, in its discussion of
Research and Development, proposes an external review group to
ensure that research is 'of benefit to the industry'. The Group,
Bell suggests, should comprise academics, staff of the Agriculture
and Food Research Council and representatives of the industry.
Conservation interests are not given specific mention. The
conclusions do not embody the emphasis found in a number of

recommendations of the House of Lords Committee (1984b) on
Agricultural and Environmental Research. That committee, for
instance, recommended that 'the Agriculture Departments should have
responsibility for promoting research on the environmental effects
of agricultural practices, whether or not such work appears likely
to have benefit in terms of farming economies'. (Vol I, 26.)

MAFF's reputation for intransigence was apparent in a survey of
national environmental groups conducted in 1980 (Lowe and Goyder
1983). Group leaders were asked whether or not they found access
easy to the relevant level of authority in the eighteen government
departments and agencies with the most extensive contacts with the
environmental lobby. They were also asked whether they found the
same departments and agencies reasonably receptive to their point of
view. MAFF came at the bottom of both lists: overall environmental
leaders regarded it as the least accessible government department
and the most unreceptive one. It is not just external groups which
have been excluded from agricultural policy making but other
government departments also. The Countryside Commission, the
government's own statutory advisor on countryside matters, was led
to complain in 1979 that:

> Over the years we have regularly had cause to regret a lack of
> regard for conservation and recreation in Ministry policy and
> practice: we have tended to make more progress dealing with the
> private farming and landowning organisations directly.
>
> (Countryside Commission, 1979).

Similarly, the Royal Commission on Environmental Pollution (1979) in
its report on Agriculture and Pollution commented 'We have formed
the view that the MAFF approach to pollution questions has been
unduly defensive and protective towards agricultural interests'.

A possible modification of the Ministry's intransigent stance
was signalled in July 1984 when, with agricultural policies
attracting intense parliamentary and press criticism, the Ministry
announced the establishment of an Environment Co-ordination Unit.
However it remains to be seen whether this will become an effective
channel for consultation with environmental interests. The
Ministry's intention seemed to be to improve its public image,
rather than to open up its policy making to outside influence. The

person appointed to head the Unit's staff of six was a food
scientist with the rank of Assistant Secretary, the fourth highest
grade within the Civil Service. As one newspaper reported at the
time 'senior ministry officials made it clear that his influence
would be limited and would be restricted to advice. He will not be
involved in policy making'. (Guardian, 24 July, 1984.)

The only major exception to the Ministry's general
unresponsiveness towards environmental groups has been its very
active relations with the Farming and Wildlife Advisory Group (Cox,
Lowe and Winter 1985). FWAG comprises a full-time officer formerly
employed by ADAS, a national committee made up of representatives
from agricultural and conservation groups and a network of county
groups several with full-time advisory officers. It has received
strong backing and encouragement from the Ministry, including
material support. Its treatment contrasts sharply with MAFF's
indifference towards other environmental groups. Perhaps this is not
surprising given that MAFF officials played a leading part in
establishing the group, along with leaders of the Royal Society for
the Protection of Birds (RSPB) and the Royal Society for Nature
Conservation (RSNC). Even so it was only with the much trumpeted
launch of the Farming and Wildlife Trust, a money raising arm for
FWAG, in February 1984 that the Ministry made a direct financial
contribution, albeit of a nominal sum, to the development of the
FWAG movement.

As a department used to corporatist and clientelist relations
with producer groups, MAFF has sought in effect to 'license' its own
environmental interest group. As such FWAG has a number of
advantages for the Ministry. The Ministry is able to counter
criticism from environmentalists by pointing to its support for FWAG
as evidence of its commitment to conservation (e.g. Wiggin 1981).
FWAG does indeed make conservation opinion readily accessible to the
Ministry and vice versa, but it does so at 'arm's length'. The
Ministry retains its independence in contrast with what might happen
if, for instance, an advisory committee were to be established. In
any case the members of FWAG are committed to seek accommodation
between agriculturalists and conservationists through discussion and

the exchange of views and information. The Group does not associate itself with pressures for legislative or policy changes.

FWAG's unifying principle is that modern, profitable farming and conservation are not incompatible and can be reconciled by providing suitable advice, information and incentives for farmers to include conservation in their farm planning and management. This particularly suits the Ministry as it avoids questions of the reform of agricultural policy or of the desirability of controls over farming practices. The Ministry has not found difficulty, therefore, in accommodating FWAG. Indeed, in many respects FWAG operates as an informal and conservation-minded version of ADAS, complementing its work and adopting similar techniques to disseminate conservation ideas, such as the use of demonstration exercises, the issuing of advisory leaflets and a focus on key progressive farmers who, it is hoped, will give a lead that others will readily follow. The final advantage of FWAG, from the Ministry's viewpoint, is that its composition defines a middle ground of moderation which marginalises certain excluded groups and issues. The conservation groups which belong to national FWAG are the more establishment orientated and conservative. The newer, radical groups such as Friends of the Earth are not included, nor are groups such as the CPRE and the Ramblers' Association which have pressed for the reform of agricultural policy or the introduction of controls over farming practices. The environmental problems of modern agriculture are also narrowed down to the issues of wildlife habitat and landscape quality. Other conflicts, such as those over pollution and access, feature minimally, if at all, in FWAG's concerns.

The Tactics of Conservation Groups

Conservation groups have not sat idly by allowing themselves to be passively manipulated by agricultural interests. One problem they face is that they possess no veto over policy making which would automatically ensure them a place in the policy community. Some major interest groups are in this position because they possess economic sanctions or their co-operation is vital to the implementation of policy or the normal functioning of government. Environmental groups lack such an effective sanction, and not being dependent upon their co-operation government is not obliged fully to accommodate them. Yet sanctions of a kind are available to environmental groups and it is possible to identify four different tactics used by environmental groups in an attempt to insinuate themselves into this and other policy communities.

These can be characterised as public censure, delay, tactical coalition and the backdoor method. Regarding the first Richardson (1977) remarks 'it is clear that many environmental groups can, under favourable conditions, generate a body of opinion to which policy makers are prepared to respond'. He suggests, however, that this weapon is common to all pressure groups and may be somewhat ephemeral in its impact. In relation to this particular case we would want to dissent from that general conclusion. The evidence of the strong links between environmental groups and the media suggests that media receptiveness is more than a passing fashion (Lowe and Morrison 1984). Moreover, other research has shown that some groups, such as trade unions, are not so favourably treated by the media (Glasgow University Media Group 1978, 1980; Hall et al 1978). If there is a deep-seated cultural bias in the operations of the media, and theories of the media need to be carefully qualified if simplistic conspiratorial accounts are to be avoided, then environmental groups seem to be among the beneficiaries.

The passage of the Wildlife and Countryside Act, in particular, set in motion a wide-ranging debate about the environmental consequences of modern agriculture. Environmental groups have used their extensive media links to fuel the debate with specific cases of habitat and landscape loss. In one parliamentary briefing the CLA complained that:

> tales of destruction in the countryside catch the headlines and are easily digested by the public mind. The quiet efforts of owners working on good conservation policies, perhaps over many years are much less sensational....The imbalance in the reporting of good conservation news leads to an unwarranted readiness by many people to accept the need for greater control.

> (CLA, 22 April, 1981).

To combat such 'tales of destruction' the CLA sought to publicise the results of its own surveys which pointed to considerable positive conservation action - planting trees, managing hedges and so forth - on the part of CLA members.

The NFU, for its part, initiated a publicity campaign with the title The Backbone of Britain, through which it hoped to 'repair damage where it has been caused between the two main sections of society - those who live and work in the town, and those who work in the country' (NFU Press Release, 29 June, 1982). It must be emphasised, of course, that the NFU is not without its media contacts. Farming, with its own daily and weekly national radio and television programmes, has a favoured position. Although such programmes are broadcast at particular offpeak times, and, by and large, reach a very specific audience, they can, by virtue of that, have important implications for the internal cohesion of the farming and landowning community. Agricultural correspondents in the national press also tend to be sympathetic to farming interests.

The way in which an issue concerning agricultural development or the countryside is handled by a newspaper seems often to depend on an editorial judgement of whether it falls within the purview of the agricultural correspondent or of the environmental or planning correspondent. Moreover, in more general terms the claims advanced by agricultural interests resonate with an oft commented upon dominant strain in national culture which has seen dependable gentlemanly virtues as rooted in the countryside (Wiener 1981). It should be added however that conservation groups also occupy part of the same tradition. Part of the ideological impact of FWAG has been in uniting agricultural and conservation interests against the 'common enemy..the urban industrial interest' (Winnifrith 1970). In the same way that the agricultural lobby has attempted to identify radical conservation opinion with urban misunderstanding so too conservationists have begun to discriminate between 'traditional' and 'industrial' farming.

Conservation groups do seem to be winning the media battle. They have been greatly helped by a new climate of opinion - encouraged by political concern over the financial cost of agricultural policy and over surplus production - that is much more sceptical towards the claims of the farming lobby than hitherto. The

ideological resources of the farming and landowning community have
suffered significant depletion. This wider media debate has in turn
sustained parliamentary interest in the issue and a string of
reports, several already referred to, has appeared from
parliamentary select committees critical of the conservation
failings of agricultural policy and of the insularity and blinkered
outlook of MAFF. One House of Lords select committee (1984a) has
condemned MAFF for being 'backward-looking', too
'production-orientated' and insufficiently responsive to the
strength of public opinion on the countryside. The committee
proposed that 'the care of the environment should have comparable
status with the production of food' in the promotion of farming
improvement, a proposal subsequently ignored by the Bell Report.

Similarly another House of Lords Committee (1984b) on
Agricultural and Environmental Research criticised the paucity of
research on the long-term environmental implications of contemporary
farming and forestry practices. A principle cause of concern was
diagnosed as 'the closed loop between the Agriculture Departments
and agricultural research', whereby research priorities were
dominated by the interests of the farming and food industries. 'The
opportunities for outside interests to contribute are heavily
circumscribed', complained the Select Committee, and 'environmental
interests are conspicuously absent from agricultural research
committees'. Finally the House of Commons Select Committee on the
Environment (1985) accused MAFF officials of 'dragging their feet'
and of being 'half-hearted' over conservation. The Ministry had not
responded adequately to the changed climate of opinion even though
'The weight of evidence that the balance between agriculture and the
environment is wrong is overwhelming'. 'MAFF must reappraise its
attitudes', urged the Committee.

The second tactic which may be available to environmental
groups excluded from the decision making process is to attempt to
block or delay the implementation of policy. Of this tactic
Richardson (1977) comments:

> The planning process, providing as it does (however
> inadequately) for some measure of public participation is by
> its nature lengthy and protracted. The opportunity to put one's

> case against a proposal may be used not only for just that, but also as a tactic.... No rational authority or company will willingly provoke amenity protests if the procedure for reaching a decision is extremely time-consuming.

The ability to inflict delay is increased not only by opportunities for participation but also through the devolved nature of statutory planning. The multiple points provided for intervention allow environmental groups to fight a rearguard action against an opposed development proposal.

As a result the agricultural lobby has seen resistance to planning regulation as an important factor in minimising the impact of environmental opposition. In other policy sectors the requirement for new developments to gain planning consent and for major and controversial proposals to be put before a public inquiry has provided an opportunity for environmental protestors to challenge policy, delay implementation, and force their way into the policy community. This has happened most notably in the transport and energy sectors. Agriculture, however, is largely not subject to planning controls. The Minister of Agriculture, at his own discretion, can call for a public inquiry to be held to examine major land drainage schemes. This was done in the case of Amberley Wildbrooks and the justification for the scheme was roundly trounced by the CPRE and RSPB. The Minister duly withdrew the scheme, but subsequently no more inquiries have been held. For example the option was steadfastly refused in relation to the controversial schemes on the Halvergate Marshes. Interestingly it was on Halvergate in 1984 that the sort of direct action tactics which appeared in other environmental conflicts in the 1970s, often linked to public inquiries, emerged for the first time in the agriculture/conservation conflict. However the potential to disrupt the implementation of agricultural policy through such tactics is limited.

The third tactic that outsider groups may pursue to ease their way into a policy community is through building a coalition with one or more of the insider groups. An example is that of Transport 2000, which was formed in 1973 to champion public transport on social and environmental grounds. It brought various national environmental

groups into direct partnership with the railway unions and the
Railway Industry Association, thereby providing the groups with a
foothold in transport policy making. Environmental groups have
pursued similar strategies in the agricultural sector. FWAG is one
outcome, though its utility to environmental groups as a bridgehead
into agricultural policy making is limited by its circumscribed
constitution which bars it from acting as a pressure group and the
tendency for the agricultural interests to be predominant on its
county committees. FWAG's whole ethos, in any case, militates
against making such headway. Composed as it is of organisations that
sometimes find themselves at odds on fundamental policy questions it
assiduously avoids adopting positions in relation to divisive issues
and carefully refrains from making pronouncements that might
embarass any of its associated organisations. Recently the Director
of the RSPB expressed his disillusionment with this approach:

> Rather than seeking to change farming policies and practice, we
> chose to become a prime mover of the Farming and Wildlife Group
> (which) sought to explore the opportunities for compromise
> between production and conservation on the farm. Latterly,
> however, it has become clear that collaboration is not enough
> - the tide of agricultural change runs so strongly that it
> threatens to overwhelm some farmland birds. But to alter
> agricultural policy is no easy matter.
>
> (Birds, Summer, 1984).

The environmental lobby has, of course, generated a number of
proposals aimed specifically at wooing sections of the agricultural
community and typified most recently by the spate of suggestions for
new types of farm grant combining income support with conservation
objectives. Such initiatives are intended particularly to attract
the support of small farmers and hill farmers (e.g. MacEwen and
Sinclair 1983). Associated with them are attempts to demonstrate
commonalities of interest hitherto only dimly perceived, if at all.
As Malcolm MacEwen puts it:

> If the farming unions would join us, the conservationists, in
> pressing for agricultural money to be used differently they
> would find themselves much less isolated than they now are.
> Farming today is dangerously exposed. It badly needs friends.
>
> (MacEwen 1984).

The notable unity of the farming lobby has generally left little
scope for the deployment of coalition tactics, not least because of

the predictable response of farmers, when challenged or courted by
external interests, to close ranks. The NFU has been almost wholly
successful in containing the divergent interests of farmers of
different regions and sectors. This has been accomplished through
its elaborate network of specialist commodity committees and county
branches, which balance and absorb internal disagreement, and
through the virtual monopoly on representing farmers' interest to
government. Thus it is difficult for those within the industry who
disagree with its policies to gain a hearing.

More recently, however, the unity of the farming lobby has been
placed under considerable strain from a number of quarters and there
are clear indications that this has helped to create a political
space within which ideas can be constructively exhanged and
realignments nurtured. The increased specialisation which has
followed EEC membership has had a profound impact on the NFU's
internal politics, and at the 1984 Annual General Meeting the unity
of the lobby appeared all too brittle as the growing rift between
arable farmers and livestock producers - popularly characterised as
'horn versus corn' - erupted in acrimonious discontent. The internal
consensus, so essential to a working partnership with the state and
relatively easily achieved in the days when mixed farming
predominated, is currently in some disarray. The Common Agricultural
Policy of the European Community has systematically favoured arable
producers (Bowers and Cheshire 1983), and the persistent failure of
EEC Agriculture Ministers to agree on proposals to bring burgeoning
cereal production under control exposes farming interests to a
public opinion outraged by the associated surpluses. Where action
has been taken, such as with the imposition of milk quotas in April
1984, the problems posed by the need to effect rapid structural
re-adjustment have brought with them significant dissent within the
Union. Such developments come, moreover, after a period which has
seen both regional voices and special interests asserting themselves
more stridently than hitherto.

Whereas separate unions in Scotland and Ulster have close links
with the NFU, which operates only in England and Wales, in Wales
competition between the NFU and the Farmers' Union of Wales (FUW) is
keen. The FUW, formed in 1955, achieved a major breakthrough in 1978

when it was granted formal recognition in the annual price review.
Since then it has had a number of issues, including several
conservation conflicts and more recently milk quotas, on which it
has been able to develop a high profile as an alternative to the NFU
at least for Welsh farmers. Then in 1981 a small number of tenant
farmers, of the opinion that the NFU was failing adequately to
represent their interests on tenancy reform, formed an Association.
A Smallfarmers' Association, set up in 1979, is another symbol of
the concern at the NFU's monopoly of representation. However none of
these developments have posed a major challenge to the NFU which
continues to speak with authority for the industry as a whole. They
do nevertheless delineate constituencies whose interests must be
accommodated as far as possible if that authority is to be retained.
Even in Wales where the challenge has been the severest, with
perhaps half of farmers belonging to the FUW, the NFU retains
considerable strength and has devoted much attention to maintaining
its presence.

In 1984, following a stormy annual meeting the NFU initiated a
wide-ranging review of its fundamental policies, christening the
operation Agriculture's Watershed. The CLA, too, undertook a policy
review, starting somewhat in advance of the NFU. The CLA focussed
particularly on the relationship between agriculture and the
environment and its Agriculture and Land Use Sub-Committee's
nineteen page report (CLA 1984) proved, in terms of the traditional
presuppositions of debate within the farming and landowning
community, to be an extraordinarily radical document. From these
developments it seems apparent that there are greater possibilities
for pursuit of the coalition tactic than at any time since the
modern era of agricultural policy was ushered in with the 1947
Agriculture Act. Thus in 1984 the CLA President joined the Chairman
of the CPRE in a public pledge that the two bodies would work
together for changes in government farm support policies to give a
better balance between agriculture and conservation. The farming
lobby no longer has the appearance of an implacable, fissureless
power block. Indeed the CLA now appears increasingly critical of the
Ministry's cautious position on conservation. In response to the
Ministry's April 1985 proposals for the reform of capital grants the

Association declared that it was an 'opportunity missed'. The proposals merely paid 'lip service to conservation' (CLA Press Release, 21 May, 1985).

The fourth tactic pursued by environmental groups is the 'backdoor method', whereby the attempt is made to make use of influence with a 'friendly' department to intervene in the affairs of an 'unfriendly' one, and so transform a previously closed policy community into an accessible 'issue network'. Much of the lobbying of environmental groups over the past few years has been concerned to press the DoE and the statutory conservation agencies to take a more active and critical stance towards agricultural policy. As the Director of the RSPB has put it, 'we are also pressing the Department of the Environment to take its own wider responsibilities more seriously and to make strenuous efforts to challenge the agriculture departments' dominance in rural affairs' (Birds, Summer 1984). It has probably always been the case that the salient constraint on the freedom of the agricultural policy community has been the concern of other government departments at the financial and other implications of developments in agricultural policy (Wilson 1977). Certainly since 1984 it has been evident that European Agricultural Ministers no longer have a free hand in dispensing largesse to the European Community's farmers, and in Britain at least it is the Treasury and the Foreign Office that have most obviously been tying the Minister's hands. Hitherto the DoE has been little involved, but with the higher political profile of environmental matters a revision of departmental relationships could have far-reaching implications.

The interaction between MAFF and the DoE came under close scrutiny from the House of Lords Select Committee (1984a) on the European Communities and examination of witnesses revealed that links between the departments were, to all intents and purposes, minimal. In response to an enquiry as to whether any recognised inter-departmental machinery existed through which the DoE and MAFF might discuss matters of common concern the Under Secretary at the DoE's Directorate of Rural Affairs had to concede that, while various ad hoc discussions took place from time to time, no institutional arrangement existed. The Committee accordingly called for the revision of existing priorities and greater co-ordination and co-operation between MAFF and the DoE. 'In the past' it complained, 'the DoE have been largely subordinate to MAFF, and have not been active enough in promoting care for the environment'.

The DoE, for their part, were urged 'to revise their role in relation to agriculture generally'.The House of Commons (1985) Committee examining the operation of the Wildlife and Countryside Act came to some very similar conclusions. While pointedly recording that 'all of our witnesses, except MAFF, recognise the need for a serious change of direction', it welcomed the setting up of MAFF's new Environmental Co-ordination Unit and other encouraging signs of developing links between MAFF and the DoE. The DoE, in fact, submitted a supplementary paper detailing the links between the departments. However, as the Committee noted, the links are 'necessarily directed externally' and certainly they indicate extensive liaison of a technical kind regarding matters of effective policy implementation but a muted voice for the DoE in policy formulation. The demand for institutionalised links to be fostered has now been firmly registered, however, and the Committee furthermore recommended that 'a working party be set up to investigate ways in which the duties and administrative structure of MAFF could incorporate a stronger conservation element in all agricultural policy'.

Clearly the sorts of developments envisaged presage the accommodation by the agricultural policy community of a more effective DoE voice in agricultural policy making which will in turn mean that the indirect 'backdoor' representation of the conservation lobby is enhanced. Already there are indications of a new assertiveness on the part of the DoE in its dealings with the Agriculture Departments, evidenced most notably in the imposition by the Secretary of State for the Environment of an Article 4 Directive to prevent the drainage of one of the holdings on the Halvergate Marshes. This move clearly challenged the claim that planning controls are totally inappropriate in regulating agricultural development. However given the resistance within departments to changes which might erode their administrative territories and associated power within the state system it remains to be seen

across how broad a front the DoE will be prepared to see itself come
into conflict with MAFF.

CONCLUSION

Standard operating procedures which underpin persistent patterns of
relationships tend to keep issues off the party political agenda and
ensure that they attract little public attention and if, as Jordan
and Richardson (1982) suggest, the dominant British policy style -
at least at the level of procedural ambition - can be characterised
in such terms as 'bureaucratic accommodation', then it is equally
evident that the agricultural policy community is now having to cope
with considerable disruption. The sector's preferred and long
regularised arrangements are being scrutinised and their rationale
questioned in a manner hitherto unprecedented in the post-war
period.

On matters relating to agricultural policy and conservation
there is an increasingly well informed and critical public opinion
and, both chastened and educated by its experience during the
passage of the Wildlife and Countryside Act 1981, the conservation
policy community has become more effective in its search for voice.
Pivotal to its efforts was the opportunity afforded by the inquiry
held in 1983-84 by the House of Lords Select Committee (1984a) and
the major pressure which was thereby put upon Agriculture and
Environment Ministers was evidenced by an Early Day Motion on the
future of the countryside which was heavily critical of present
agricultural policy and was signed by nearly one-third of all MPs,
half of them Conservative. With some farmers themselves coming to
share the concerns of the environmental lobby the agriculture and
conservation issue is now firmly on the main political agenda, with
no prospect of its being marginalised again.

The reports of the recent parliamentary select committees, the
public statements of the statutory conservation agencies and other
organisations with interests in the countryside,and media coverage
are all indicative of wide political, institutional and public
support for change. And given that policy making in Britain is
typically reactive and that the agricultural policy community has

been a notably closed one the environmental lobby can now count some significant - albeit interim - successes, if only because their efforts have compelled a series of reactions. Whether such gains can be translated into an enhanced voice in relation to the broader aspects of agricultural policy making is much less clear, however. Certainly the ambition to radically transform the inner-circle negotiation which has hitherto identified the agricultural policy sector has been clearly stated. Speaking at a Cirencester agricultural conference in January 1985 Robin Grove-White claimed that 'bodies like the RSPB, Friends of the Earth and my own CPRE are now no longer content to leave agricultural policy to be determined in cabal by MAFF, the NFU and the CLA' and he went on speculatively to envisage that environmental groups would be 'seeking and getting' a say in how patterns of grant aid and price support are determined in both Whitehall and Brussels (Daily Telegraph, 5 January, 1985).

But while it is clear that a political space has been opened up within which re-adjustments of position can take place consideration of the resources which can be mobilised by contending groups suggest that their scope may be predictably constrained. For environmental groups declared intention and prospective achievement remain severely attenuated. MAFF in particular is working hard to sustain the appearance of accommodation while it seeks to consolidate the emergent consensus around the voluntary philosophy embodied in the Wildlife and Countryside Act. The repeated chronicling of efforts actively to recognise conservation interest corresponds well enough to the processes of absorption that Wildavsky (1975) noted in British policy making but the uncompromising response both to the Clarke Amendment Bill and to the House of Commons Select Committee report on the operation and effectiveness of the 1981 Act (e.g. MAFF, Press Release, 23 May, 1985) hardly manifests the anticipatory adjustments that he saw as accompanying and making possible such processes. The MAFF strategy, rather, has been one of resolute rearguard action orientated towards the maintenance of order within the agricultural policy sector. The DoE meanwhile, notwithstanding its new assertiveness, is showing no obvious relish for confrontation with MAFF over major issues.

So although recent developments have unquestionably driven a

number of wedges into the agricultural policy community there is
little indication as yet that their cumulative effect has been to
prise it apart. Insofar as it currently has an appearance of
unprecedented receptiveness the explanation is more appropriately to
be found in shrewd calculative responses to the range of pressures
- notably from Europe - rather than a fundamental re-orientation of
policy philosophy. Indeed, in the face of articulate demands for
radical change there is an obvious determination to ensure that the
prevailing orthodoxy be modified only in an incremental manner
through a series of piecemeal accommodations. The MAFF strategy
still perfectly exemplifies Paul Valery's dictum that 'politics is
the art of preventing people from taking part in affairs that
properly concern them'.

To conceptualise the state as a unitary structure would be a
dereliction of analytic duty since adequately understanding the
politics of the policy process minimally involves identifying
relevant policy sectors and their associated policy communities.
Consideration of the normalised pattern of their relationships and
their reactions when procedures are placed under stress demonstrates
just how inappropriate it is to see pressure groups, for instance,
as simply lobbying the centre. Some groups have succeeded in
colonising parts of the centre and have become themselves
incorporated by the centre as part of that self-same process. The
state structures which emerge are characterised in turn by a
departmental pluralism which finds government departments adopting
an advocacy role in relation to client groups, mobilising their
support in the inter-departmental competition for administrative
responsibilities, legislative time and resources.

In such a brokerage system the mythology of the unitary state
can be both invoked defensively to mediate contradictions and
deployed critically to expose them. With the agricultural policy
sector increasingly beleaguered by attacks on its budgetary
extravagance and environmental insensitivity the terms of the debate
concerning agriculture and conservation are being broadened by both
relevant policy communities to include yet wider issues of rural
prosperity and the viability of rural communities. With the issue
agenda thus extended issue networks are likely to become more

fragmented and departments will find it harder to indulge their prevalent tendency to resist changes which might modify their power. The exasperated response of the House of Commons Environment Committee to a statement of departmental demarcations from MAFF may yet prove to have a resonance beyond the issue of capital grants which was its immediate pretext:

> The Crown is one and indivisible: at least in constitutional theory. It is absurd that departments of the Crown should not be pursuing policies and applying public monies in a manner which is consistent and complementary with one another.
>
> (House of Commons Environment Committee, 1985).

FOOTNOTES

1 This paper arises out of a project on land use conflict after the Wildlife and Countryside Act financed by the ESRC. We gratefully acknowledge their support.

2 For a full analysis of these groups see Lowe and Goyder (1983).

3 The distinction between 'wildlife' and 'countryside' concerns is a peculiarly British phenomenon exemplified in the existence of separate government agencies. While the NCC exists to further the interests of wildlife and nature conservation the Countryside Commission has a wider remit which includes landscape protection and the promotion of countryside recreation. There is now a growing body of conservationist opinion which deplores such a division and seeks the amalgamation of the two bodies.

4 For an analysis of a Rayner review in a different sphere of government see Warner (1984).

5 The distinction between corporatism and clientelism needs to be made here. Jessop (1982) defines clientelism 'as a form of representation based on the exchange of political support in return for the allocation of politically-mediated resources and involving a hierarchical relationship between dependent client(s) and superordinate patron(s)' and corporatism as 'representation on the basis of function within the division of labour.. characterised by the formal equivalence of 'corporations''. Clearly the two forms of representation can be closely linked and the concepts do not describe political arrangements which are mutually incompatible. By the same token we do not use the terms in the way that they are sometimes, at a more abstract level of theorising, to characterise specific polities or state forms (cf. Dyson 1980). See also Jessop (1979).

6 A government amendment to the Agriculture Bill, along the lines of the original Clarke proposals, was tabled in January, 1986, representing a shift from the position described here.

7 We are currently researching the role of the Farming and Wildlife Advisory Group which, amongst other things, seeks to liaise between different bodies with an interest in conservation and co-ordinate the conservation advice available to farmers. For some preliminary thoughts on its role see Cox, Lowe and Winter (1985).

8 ADAS is divided into four services - Agriculture, Agricultural Science, Land and Water, and Veterinary. The Agriculture and Veterinary are the largest two services, respectively providing agricultural production advice to farmers and controlling, eradicating and investigating animal diseases. LAWS is concerned with capital investment in infrastructure and the use of land as a resource. It has a tradition of employing broadly trained land managers and land surveyors often more sympathetic to the principles of conservation than the more narrowly production orientated staff of the Agriculture Service.

REFERENCES

Bell, R. 1984. Report of a study of ADAS. London: HMSO.

Bowers, J. and Cheshire, P. 1983. Agriculture, the countryside and land use. London: Methuen.

Brotherton, I. and Lowe, P. 1984. Statutory bodies and rural conservation: agency or instrument? Land Use Policy 1, 147-153.

Cawson, A. and Saunders, P. 1983. Corporatism, competitive politics and class struggle. In Capital and politics, R. King (ed), 8-27. London: Routledge and Kegan Paul.

CLA 1984. Advisory group on the integration of agricultural and environmental policies: report and recommendations, Press Release: Sept 24.

Cox, G. and Lowe, P. 1983a. A battle not the war: the politics of the Wildlife and Countryside Act 1981. Countryside Planning Yearbook 4, 48-76.

Cox, G. and Lowe, P. 1983b. Countryside politics: goodbye to goodwill? Pol. Quarterly 54, 268-282.

Cox, G. and Lowe, P. 1984. Agricultural corporatism and rural conservation. In Locality and rurality, T. Bradley and P. Lowe (eds), 147-166. Norwich: Geo Books.

Cox, G., Lowe, P. and Winter, M. 1985. Land use conflict after the Wildlife and Countryside Act, 1981: the role of the Farming and Wildlife Advisory Group. J. Rural Stud. 1, 173-183.

Countryside Commission 1979. Annual report. Cheltenham.

Dyson, K. 1984. The state tradition in Western Europe. Oxford: Martin Robertson.

Glasgow University Media Group 1978. Bad news. London: Routledge and Kegan Paul.

Glasgow University Media Group 1980. More bad news. London: Routledge and Kegan Paul.

Grant, W. The National Farmers' Union: the classic case of incorporation? In Pressure politics, D. Marsh (ed), 129-143. London: Junction Books.

Hall, S., Critcher, C., Jefferson, T., Clarke, J. and Roberts, B. 1978. Policing the crisis: mugging, the state and law and order. London: Macmillan.

Heclo, H. 1978. Issue networks and the executive establishment. In The new American politics, A. King (ed). Washington: American Enterprise Institute.

House of Commons Environment Committee 1985. Operation and effectiveness of part II of the Wildlife and Countryside Act. Session 1984-85, 1st Report. HC 6-11.

House of Lords Select Committee on the European Communities 1984a. Agriculture and the environment. Session 1983-84, 20th Report. HL 247.

House of Lords Select Committee on Science and Technology 1984b. Agricultural and environmental research. Session 1983-84, 4th Report. HL 272.

Jessop, B. 1979. Corporatism, parliamentarism and social democracy. In Trends towards corporatist intermediation, P.C. Schmitter and G. Lehmbruch (eds), 185-212. London and Beverley Hills: Sage.

Jessop, B. 1982. The capitalist state. Oxford: Martin Robertson.

Jordan, A. G. 1981. Iron triangles, woolly corporatism and elastic nets: images of the policy process. J. Public Policy 1, 95-123.

Jordan, A. G. and Richardson, J. J. 1982. The British policy style
    or the logic of negotiation? In Policy styles in Western
    Europe, J. J. Richardson (ed), 80-110. London: George Allen and
    Unwin.
Lowe, P. and Goyder, J. 1983. Environmental groups in politics.
    London: George Allen and Unwin.
Lowe, P. and Morrison, D. 1984. Bad news or good news: environmental
    politics and the mass media. Sociological Review 32, 75-90.
Lowe, P., Cox, G., MacEwen, M., O'Riordian, T. and Winter, M. 1986.
    Countryside conflicts: the politics of farming, forestry and
    conservation. London: Gower and Maurice Temple Smith.,
MacEwen, M. 1984. Agriculture and conservation. Ecos: A Review of
    Conservation 5, 4, 26-30.
MacEwen, M. and Sinclair, G. 1983. New life for the hills. London:
    Council for National Parks.
MAFF Advisory Council for Agriculture and Horticulture 1978.
    Agriculture and the countryside. London: HMSO.
Marsh, D. (ed) 1983. Pressure politics. London: Junction Books.
Metcalfe, L. and McQuillan, M. 1979. Corporatism or industrial
    democracy? Pol. Stud. 27, 266-282.
Middlemas, K. 1980. Politics in industrial society: the experience
    of the British system since 1911. London: Andre Deutsch.
Newby, H. 1979. Green and pleasant land? social change in rural
    England. London: Hutchinson.
Richardson, J. J. 1977. The environmental issue and the public. In
    Decision making in Britain: pollution and the environment.
    Milton Keynes: Open University Press.
Richardson, J. J. and Jordan, A. G. 1979. Governing under pressure.
    Oxford: Martin Robertson.
Richardson, J. J., Jordan, A. G. and Kimber, R. H. 1978. Lobbying,
    administrative reform and policy styles. Pol. Stud. 26, 47-60.
Royal Commission on Environmental Pollution, 1979. Agriculture and
    pollution. London: HMSO.
Self, P. and Storing, H. 1962. The state and the farmer. London:
    George Allen and Unwin.
Warner, N. 1984. Raynerism in practice: anatomy of a Rayner
    scrutiny. Public Admin. 62, 7-22.
Wiggin, J. 1981. Agriculture and conservation. Ecos: A Review of
    Conservation 2, 8-9.
Wiener, M. J. 1981. English culture and the decline of the
    industrial spirit, 1850-1950. Cambridge: Cambridge University
    Press.
Wildavsky, A. 1975. Budgeting: a comparative theory of budgeting
    processes. Boston Mass: Little Brown.
Wilkinson, J. M. (ed) 1983. Responsible agriculture. Rural: Report
    of the Inaugral Conference.
Wilson, G. K. 1977. Special interests and policy making. Chichester:
    John Wiley.
Winnifrith, J. 1970. The clash between conservationists and farmers.
    In Farming and wildlife: a study in compromise, D. Barber (ed),
    14-21. Sandy: RSPB.
Wootton, G. 1978. Pressure politics in contemporary Britain.
    Lexington, Mass: Lexington Books.

12 Agricultural policy and party politics in post-war Britain

ANDREW FLYNN

INTRODUCTION

This chapter analyses the post-war development of Labour and
Conservative agricultural policies and assesses the utility of
models of party competition in explaining the course which policies
took. As Parkinson (1970) points out, party activity has to some
extent been neglected in contemporary research on British politics.
This is largely due to the widespread belief that national parties
are of only peripheral importance in today's international,
interdependent economies. Yet political parties can be important in
defining, formulating and to some extent implementing policies.[1]
Moreover they can influence the climate of opinion within society,
which allows some forms of action to be considered in a more or less
favourable light, and a number of commentators have identified party
competition as the dominant factor in agricultural policy formation.

THEORIES OF PARTY COMPETITION

Crudely, the argument goes that, in order to court the rural vote,
the major parties have competed with one another to present
favourable agricultural policies, leading to a convergence of party
positions. Self and Storing (1962) put forward such an argument,
claiming that the passage of the 1947 Agriculture Act by a Labour
Government brought with it the advent of party competition in
agricultural policy-making. With its new found rural popularity
Labour was trying to change what were believed to be farmers'

traditional political alliances. This also involved the shelving of
policies which would be electorally unpopular with farmers, such as
land nationalisation. Labour's vigorous campaign for the farming
vote demanded a response from the Conservatives to renew their
agricultural appeal. As Self and Storing conclude:

> In the last 15 years (1947-1962) politicians seem mainly to
> have wanted to outdo one another in their favours to
> agriculture. It is true that their ardour was keenest when the
> responsibilities of office were most remote, but all the same
> agriculture has come a long way since the time when it
> considered itself ignored by both main parties, the one
> confident that it had nothing to lose the other sure it had
> nothing to gain.
>
> (Self and Storing 1962).

The parties were competing for the votes of a section of the
electorate and were adapting their policies accordingly - there
appeared to be a large area of agreement emerging. Policy was being
placed within bounds to satisfy the agricultural interest groups and
the rural electorate.

Gardner (1979) has gone on to suggest that, on agricultural
policy, little has separated the two major parties during the
post-war period. From a review of party manifestos, he concludes
that:

> Since the war...agriculture has been a politically quiet area
> and the extent to which it has been a non-controversial area is
> shown by the common elements in party literature. Both parties
> regularly promise support for healthy agriculture and better
> marketing; generally they have expressed themselves in favour
> of continuing the Annual Price Review and of improving
> conditions for agricultural workers. Both have at times
> favoured the provision of better medium term credit.... The
> bulk of the post-war period may reasonably be described as a
> time when, if there was not an agreed bi-partisan policy for
> agriculture, there was no general conflict between the main
> political parties.
>
> (Gardner 1979).

Gardner's analysis and characterisation, however, are subject
to a number of difficulties. Manifestos contain a series of
generalities with which to bolster party activists and appeal to the
electorate. They are not particularly useful documents through which
to establish a party's position on a particular topic or to
demonstrate similarities and differences between party policies.

Gardner argues that the manifestos for 1966 were particularly close in their agricultural policies, but this judgement does not accord with the parties' own perceptions at the time. Indeed, the Conservatives were to proclaim that 'for the first time in over a decade there is a clear distinction between the agricultural policies of the two major parties' (Conservative Party 1966).

To establish that a consensus exists between the parties one should address the means as well as the stated ends of policy. This is a critical problem for those who perceive a convergence. It would surely be politically unwise for either major party to disagree with such bland aims as support for a healthy agriculture or that of improving the conditions of the farmworker. Such statements reveal little about how these broad aims would be achieved, and if they are accepted at face value can give the false impression of harmony. Indeed, Gardner has to admit that there have been important Agriculture Acts which disturb the picture of apparent consensus construed from successive manifestos.

The model of party competition employed by Self and Storing and Gardner is extremely crude. Political parties, it is implied, are able to take up any policy position if it will gain them votes. However, parties carry with them certain traditions and beliefs which restrict their freedom of manoeuvre. Further this simplistic argument has nothing to say on the rationality of voters and politicians or the ease with which they can or will change their political allegiance.

In recent years there have been significant developments in conceptualising party competition. Potentially these theories now provide explanations of why political parties adopt certain issues or policies. By utilising these theories an analysis can be made of the development of post-war agricultural policy within the political parties. It should then be possible to assess the credibility of theories of party competition and their applicability to agricultural policy.

In his book An Economic Theory of Democracy Anthony Downs (1957) first developed a rational choice model of voters and parties. This pioneering work has proved extremely influential, providing for example the basis of Gardner's study, but it does have a number of fatal, empirical and theoretical flaws.[2] Recently, however, Robertson (1976) and Budge and Farlie (1977) have overcome many of the problems that confronted Downs, and have developed theoretically sophisticated and empirically testable models of party competition that do not assume such things as perfect information, perfect knowledge or voter rationality. Budge and Farlie have applied their model to ten countries with what they judge to be highly acceptable results. For our purposes it is only necessary to consider part of what are complex, and in Budge and Farlie's case, wide ranging models covering the behaviour of party leaders, candidates, activists and voters.

Robertson's central concept is that of a policy space.[3] He recognises that not all positions within a policy space are open to all parties equally; certain positions are occupied exclusively by one party or another. For example Conservatives could not claim to be more in favour of the National Health Service than Labour as this would create disbelief and loss of credibility among their members and loyal voters. Instead a party prefers to de-emphasise such issues and stress themes which are more favourable to itself. Thus there are limits on the extent of possible policy changes. Further parties also maintain blocks of traditional support which they do not wish to alienate and which can provide further limitations on the extent of a party's movement. Within this framework we can examine the central prediction that the Labour and Conservative parties will converge in their agricultural policies when they are both competing for the rural vote, and diverge when one party is recognised to be dominant.

THE PARTIES AND AGRICULTURAL POLICY

Post-war Labour Party agricultural policy was conditioned by three factors. Briefly these were, first, Labour's campaign to win rural seats which had begun in the mid 1930s and continued after World War II, and which was regarded by the party as a success.[4] Labour was doing better than ever before in the rural constituencies. This campaign also entailed some toning down of the more controversial aspects of Labour's rural policy, such as land nationalisation, as

the party attempted to compete for the rural vote.[5] Secondly, the
crisis in agriculture in the inter-war years and the resultant rural
poverty convinced the party that full, planned production was
essential. Thirdly, the imposition of agricultural controls and
guarantees in both World Wars highlighted the benefits of planned
agricultural production.

With the election of a Labour Government the potential for
agricultural competition now existed and we would expect a
coincidence between the parties in this area of policy. In fact
there did appear to be a substantial amount of common ground between
the political parties and interest groups. With the outbreak of war
the Government gave pledges to provide long term support to British
agriculture by guaranteeing a system of fixed prices and assured
markets, a policy which was later endorsed by the Labour and Liberal
Parties. The Conservatives even looked to the possibility of a pact
between the parties on agricultural policy, claiming that if

> by consultation between parties, certain underlying ideas could
> be agreed; if the principle could be established that without
> due cause, and without due notice, profound changes in
> Agricultural Policy should not be introduced, the resulting
> benefit to the Industry would be impossible to exaggerate.
>
> (Conservative Party 1943).

This explicit acceptance of the ideal of party unity in
agricultural policy was regarded as a major triumph by agricultural
organisations. The National Farmers' Union (NFU) commented:

> For many years before this war farmers aspired to getting
> agriculture out of 'the rut of Party politics' so that an
> agreed policy could be adopted which would give their industry
> the measure of stability which it so sadly needed. Now the
> three major political parties have jointly undertaken to place
> on the statute book just such a policy. That means that today
> there is a general recognition of the fact that the national
> interest demands that there shall be no repetition of the
> betrayal which followed the War of 1914-18.
>
> (NFU 1943).

The Royal Agricultural Society of England too supported the
Government's policy and argued that

> ...no matter how carefully the details of any policy are
> planned, the agricultural industry can never have the necessary
> confidence in the stability of such policy unless the industry

is removed as far as possible from the vicissitudes of party
controversy.

(RASE 1941).

In such a favoured position the role of interest groups can be that
much easier and more influential. With the principles of a policy
favourable to farmers settled, changes of government would be of
minor concern. It is the details and implementation of policy which
would become important, an area in which interest groups can play a
central role.

The first test for this new trend of convergence was the 1947
Agriculture Act. This was agreed to by both major parties and the
relevant interest groups, but beneath the surface unity there was
substantial disagreement. The Act itself was largely a continuation
of war-time measures and most of the controversy centered on
interpretations of the necessary size of the home market for
agricultural produce. In the White Paper explaining the Bill it was
stated that agriculture was to produce as much food as was desired
in the national interest. This objective, however, appeared to imply
that it might at times be necessary to limit the size of home
production.

The NFU was well aware of such implications but decided to
support the measure believing that it would not be possible to
strengthen this section of the Bill and furthermore that any attempt
would cause unwanted disagreement with the Government. This was
against the wishes of many NFU branches, and their dissent was
picked up by the Conservative Party. During the Bill's passage
through Parliament Opposition speakers argued that because the Bill
failed to specify the size of the market to be guaranteed it did not
provide any real assurance of long term stability. The Conservative
Party's position was summed up in a note prepared by its research
staff:

> The enthusiasm for the Bill expressed by NFU Headquarters is
> not shared by many county branches, and there is a general
> feeling that present prices do not justify the 'all out' effort
> which must be made, if the country's food supplies are to be
> maintained or increased. With no long-term guarantee for the
> future the incentive to increase production now is lacking....
> On these grounds it should not be difficult to establish a

sound argument for opposing the Bill should the Party decide to do so.

(Conservative Party 1947).

Thus the putative post-war agreement of agriculture almost foundered at the first hurdle. A paradoxical situation arose wherein the Conservative Party which traditionally allied itself with the farmers and the countryside was prepared to endanger the post-war strategy of the farmers' interest group, the NFU. In the event, even though they had political and policy reasons for opposing the Bill, the Shadow Cabinet wisely decided not to do so. The Labour Party would have been able to make a great deal of political capital out of Conservative opposition to an Act which so favoured farmers. Even so, given Labour's electoral success in the countryside, the Conservative Party could not afford positively and wholeheartedly to participate in the passage of a popular Labour Party measure. There is a subtle but important distinction at this time between public support and private manoeuvre in response to Labour's policy initiative, as the Conservative Party strove to keep agriculture within the political arena. This somewhat unresolved situation was to continue the next year with the publication of the Conservative Party's policy statement, the 'Agricultural Charter'.

As in the party's criticism of Labour's 1947 Act the main focus of the Charter was upon promises over the size of the market for home producers. There were, of course, a number of differences on points of detail or stance between the parties. These were principally Conservative objections to the extent and nature of controls introduced in the 1947 Act (Conservative Party 1947). At a press conference to launch the Agricultural Charter, however, R.A. Butler was unable to point to any fundamental differences between it and Government policy (Hoffman 1964).

Meanwhile Labour Party policy was increasingly orientated towards the objective of cheap food. In 1945 the Government's policy had been one of low cost production (Labour Party 1946a), by 1947 rising costs were accepted in an effort to encourage farmers' to raise production, but by 1948 the aim was once again cheap food (Labour Party 1948). This objective was seen to be essential for a large working class, urban population, and it also reflected the

strength of urban interests within the party. The policy was also
farmer orientated as farmers were the main beneficiaries of an
expanding agriculture, and Labour readily posed as the farmers'
friend. The Tories could not be trusted to maintain a prosperous
countryside, Labour claimed, as the inter-war years had illustrated.

For the farmworker, though, it was a different situation. There
was little sustained furthering of his interests either in relation
to wages or the tied cottage. The farmworkers' union, the NUAW, was
to complain to the party:

> That they were deeply concerned with the failure of the Labour
> Party and the Minister of Agriculture to implement the Labour
> Party's policy of lifting up the standards of employment of the
> workers on the land to those of the skilled workers in other
> industries.
>
> (Labour Party 1946b).

The other issue which greatly concerned the Union was the tied
cottage. Despite a party pledge to abolish it, the Labour Government
came to the view that the only way to solve the tied cottage problem
was to build more houses (Labour Party 1950). In fact the one
concrete commitment given, in a speech by Hugh Dalton in 1951, was
to put an end to the cottage certificate system which allowed for
the creation of new tied cottages.

Once post-war agricultural policy had been established there
was a tendency towards inertia. Far-reaching change was difficult
for a number of reasons. First, and very importantly, the policy
which had been initiated by the Labour Government and was
essentially maintained by the Tories after they won the 1951 General
Election was widely regarded as successful. Secondly, it was a
popular policy especially amongst the farming community. Thirdly,
any major policy alterations could have had damaging political
implications. Post-war agricultural policy was a publicly agreed one
between the major parties and farmers' interest groups; any change
in policy not sanctioned by the interest groups could well have led
to accusations of betrayal.

The situation which had emerged was somewhat confused. As Self
and Storing (1962) assert the 1947 Act did bring with it party
competition. Yet this was tempered by the actions of the

agricultural interest groups who wished to remove agricultural
policy from the arena of party politics, and who did influence the
shape of policy and the parties' stances. Another constraint imposed
on the positions the parties could adopt was their method of
policy-making. Conservative policy was controlled by the party
leadership. As Hoffman (1964) shows, the Agricultural Charter was
the work of a few individuals, despite wide ranging consultations
and internal party pressures for broader participation. Thus the
Conservative leadership was able to point policy in the directions
which it wanted and which suited its short-term political
objectives, but the Labour Party's more democratic, participative
structure was dominated by the interests of the urban working class.
These are constraints which are overlooked by those who adopt a
thoroughgoing party competition perspective.

With the Conservatives regaining their political hold upon the
countryside, according to the theory of party competition one would
expect the Labour Party either to adopt policies which would satisfy
its 'natural' supporters in the countryside, i.e. the rural working
class, or else to adopt policies very similar to those of its
successful political opponents. In the event, the Labour Party
neither pursued the trends in Conservative policy, nor established
policies which explicitly favoured the rural working class. After
its election defeat, the Labour Party National Executive Committee
(NEC) established an Agriculture, Food and Rural Life Sub-Committee
in early 1952 to investigate production and distribution matters. In
its report to the Policy Committee of the NEC the sub-committee
argued that party policy should continue to be based on a system of
guaranteed prices and assured markets as 'this is the best way to
give security and incentives to farmers'. The proposal was repeated
in the sub-committee's Interim Report which became part of
'Challenge to Britain', a policy statement published in June 1953.
This document was accepted by the 1953 party conference, formally
the party's supreme policy-making body. Even in the late 1950s the
party remained committed to the 1947 Act. At the 1956 party
conference a successful resolution called for the full
implementation of the 1947 Act. Tom Williams MP, the Act's
architect, strongly supported this position and stated:

> One thing I would prevail upon the (National) Executive
> (Committee) not to do between now and the time of the next
> Election is to start trying to find new policies for
> agriculture...we do not want another policy...the 1947 Act...
> (is) one of the most constructive pieces of legislation that
> has been put on the Statute Book of this country.
>
> (Labour Party 1956).

Support for the principles of the 1947 Act was contained in the
party's agricultural policy statement 'Prosper the Plough' issued in
1958. Again this policy document was accepted by the party
conference which once more proclaimed its continued support for the
Act.

EEC ENTRY AND PARTY COMPETITION

It was at this time that the issue of entry to the EEC increasingly
came to engage the party's attention. The main problem with entry to
the EEC was not perceived to be protecting agriculture but
containing undesirable rises in food prices (Labour Party 1961). To
keep abreast of developments within agriculture a small committee of
senior figures was established. Yet the draft report which the
committee produced, rather strangely, contained no mention of the
EEC (Labour Party 1963a). Apparently the issue was already becoming
fudged: by avoiding discussion of the EEC, divisions within the
party were allowed to lie dormant. There were, though, innovations
in the report. To maintain supplies from overseas and stabilise
world prices the report proposed to conclude wherever possible
international commodity agreements. Meanwhile guaranteed prices
would continue to exist to ensure efficient production and maintain
the confidence of farmers. The report was presented to the 1963
party conference as 'Labour and the Rural Areas' by Richard Crossman
MP who claimed that it was 'the first really new agricultural policy
we have put to the nation since we left office'. (Labour Party
1963b). Although highlighting the inertia in Labour policy-making in
the meantime Crossman does exaggerate the extent of change. The
policy was still firmly rooted in the 1947 Act.

Perhaps surprisingly (especially if one accepts a model of
party competition), it was the Conservatives who were the party of
change. For the Conservatives, the return to dominance in the

countryside should have made them content to maintain a policy that
satisfied farmers and therefore secured their electoral base. In
other words, one would expect existing policies to continue with
only marginal changes. When they returned to power the aim of the
Conservative Government was 'to enable the industry to go forward
and increase production...in conditions of greater freedom'
(Conservative Party 1953). It was only with the ending of post-war
scarcity that controls could be practicably replaced by a
Conservative Government with a philosophy of decontrol, economic
freedom and competition. The point was clearly expressed by
Churchill in the House of Commons in November 1953:

> It is our theme and policy to reduce controls and restrictions
> as much as possible and to reverse, if not to abolish, the
> tendency to State purchase and marketing which is a
> characteristic of the Socialist philosophy. We hope instead to
> develop individual enterprise founded in the main on the laws
> of supply and demand and to restore to the interchange of goods
> and services that variety, flexibility and ingenuity and
> incentive on which we believe the fertility and liveliness of
> economic life depend.... In the agricultural field, however,
> another set of arguments must be borne in mind.... It is not an
> easy task to reconcile the beneficial liberation of our food
> supply from Government controls with that effective stimulus of
> home production which is essential.... It is necessary for the
> Exchequer to subsidise in one form or another, so as to bridge
> the gap between the price level reached in a free market on the
> one hand, and the price level necessary to sustain the welfare
> of the farmers on the other.

Clearly there was a contradiction over a fundamental element of
Conservative policy: there was to be widespread decontrol and
competition within the economy except for agriculture. Agriculture
with probably as many controls as any other industry was not to have
them all withdrawn, by a long way. Instead there would be partial
reforms, some controls based upon the 1947 Act would remain, but
limits were to be placed on the extent of decontrol and
deregulation.

That there was only partial implementation of Conservative
economic philosophy within agriculture was largely due to the
strength of the farming lobby within the party. Bureaucratic
controls and sanctions over producers could be removed with the
support of the whole party. To go further and reform the system of
state supported and regulated agricultural policy would alienate the

farming section of the party. This well organised and vocal
sectional interest showed its influence at party conferences, for
example in 1953, when delegates warned of the possible dangers of
decontrol. Although conference has very little positive power in
initiating or formulating policy, it was in this instance trying to
influence the direction of policy in a negative manner by limiting
the extent of economic liberalism. Further, the party's
Parliamentary Agriculture Committee was also concerned about the
effects of decontrol (Conservative Party 1954). This committee,
which appears to have been far more influential than many of its
counterparts, showed that there were strict limits to the extent of
decontrol which the farming section of the party was prepared to
accept. Finally, Tory paternalists tended to associate themselves
with the countryside and such people were by no means as averse as
their more laissez-faire brethren to state intervention within the
economy in general and agriculture in particular.

Yet while in the early 1950s there were limits to the extent of
decontrol and reform, new factors were emerging nationally and
internationally which would precipitate further changes in policy.
First, the cost of support for agriculture was rising excessively,
which combined with concern over inflation within the economy, led
the Government to modify its objectives. The costs of production
were becoming ever more important and as a result the Government
began increasingly to emphasise the benefits of economic, efficient
production (Conservative Party 1955, 1957a), and refused to raise
guarantees as much as farmers would have wished. An attempt to
boost economic production and improve farmers' confidence for the
future was provided by the 1957 Agriculture Act which imposed
limits on the possible annual reductions in the level of guarantees.
Secondly, from the mid 1950s, problems over the rising cost of
agricultural support were further complicated by the creation of the
EEC. The Government was well aware of the political and economic
potential of the EEC and, by late 1957, negotiations with the EEC
were well under way. Part of the negotiations involved the structure
of agricultural policy but initially at least agriculture was not
accorded a significant role (Conservative Party 1957b). Backbench
Conservative MPs, though, were not prepared to accept the

possibility of agriculture being sacrificed in any negotiations.
Their views were made plain to Ministers at a series of meetings of
the party's Agriculture, Fisheries and Food Committee and through an
exchange of letters with party chairman, R.A. Butler, which pointed
to the benefits of the existing system of support and the political
dangers to the party if agriculture should suffer depression within
a European system.

With the failure of Britain's EEC negotiations and the
formation of the European Free Trade Association (EFTA) it appeared
that the defenders of the status quo had successfully resisted
pressures for change. Agricultural products were excluded from the
EFTA agreement, apart from a bilateral agreement between Britain and
Denmark, which it was claimed would not affect the Agriculture Acts
and would have minimal effects on British producers. It is
interesting to note that in the Government's explanation for the
collapse of the EEC talks, agriculture was singled out as 'one of
the reasons why we did not conclude an agreement with the Common
Market countries, and we helped to sponsor the European Free Trade
Association in which our agricultural commitments were strictly
limited' (Conservative Party 1960).

Even so it was only a temporary respite, and overtures to the
EEC were again made in the early 1960s. Moreover the cost of
subsidies continued to rise to what the party considered to be
unacceptable levels. Changes to limit the amount of direct Exchequer
payments were felt to be necessary, especially as the cost of
support for agriculture was highlighted in the Budget in a way which
was not applicable to industries protected by tariffs or other
means. Tariff protection was the method adopted by the EEC and meant
a shift in the means of support from taxpayer to consumer, and an
increase in food bills. Such a development although not likely to be
electorally popular (rising food prices never are) could easily be
accommodated within the trends in Conservative thought. A move
towards tariff support could be portrayed as a move to an internal
market economy which would help to restore consumer sovereignty and
lessen the economically distorting influences of government
intervention. It would also make future negotiations to enter the
EEC easier. By 1962 the party propaganda machinery was fully behind

entry to the EEC, and, therefore, a change in the means of agricultural support which this would entail.

However Britain's negotiations with the EEC collapsed in January 1963 and with pressures on the support system continuing, the Government was in a difficult position. It was realised that changes were necessary, but at the same time the party had to retain farmers' confidence and, with a general election looming in 1964, the rural vote.[6] The result was an attempt to limit Exchequer liability by restricting imports while publicly remaining committed to the 1947 and 1957 Agriculture Acts.[7]

By the time of the Heath Government in 1970, policy was firmly based upon the replacement of deficiency payments by food levies upon imports (Godber 1970). Yet these changes in Conservative policy are not anticipated by a party competition perspective. Admittedly the policies of the parties diverge as competition for the rural vote subsides, as the theory would predict, but it is the wrong party which moves. Thus, while the theory can appear correct at the macro level it fails to show which party moves in policy terms and more importantly why.

While the basis of Conservative policy changed markedly, so Labour's remained static and backward looking. Deficiency payments continued to be favoured throughout the period of Labour Government 1964-70 both for the security they were believed to offer the farmer and to keep prices low for the consumer. After the failure of the Labour Government's negotiations with the EEC the party continued to be overwhelmingly concerned in its agricultural policy with the farmer and consumer to the neglect of the farmworker (Robins 1979, Labour Party 1970). When the party returned to opposition there was little change in this stance. The potential effects of the Common Agricultural Policy (CAP) were analysed exclusively in terms of food prices, reflecting the party's traditional concern with the urban consumer rather than with the welfare of the farmworker or the development of rural Britain (Labour Party 1971). In a draft report by the Agriculture Sub-Committee of March 1972 it was argued that the party, when returned to power, should seek immediate price cuts and pursue a long term policy of changing CAP to something more nearly resembling Britain's previous system of subsidies for agriculture, which would enable cheap food to be available for consumers. This draft policy was incorporated into 'Priorities in Government, Labour's Aims for Britain' which was presented to the 1972 party conference as a first draft and in a final form as 'Labour's Programme 1973' to the 1973 party conference.

The Labour Party was obviously not prepared to accept the existing structure of CAP and on returning to power in 1974 already had a series of points which it wanted renegotiated. In its attempt to reform CAP, the Labour Government claimed it had been largely successful (Labour Party 1975), but the party as a whole remained unconvinced. At a special conference in 1975 the party voted against continued membership of the EEC. In a move to stifle intra-party divisions a referendum was called to decide whether or not Britain should remain a member of the EEC. Unfortunately for the mass party which had voted to withdraw, the nation decided that it wanted to remain in membership. Labour's agricultural policy now adopted a two pronged approach. First, home production was to be expanded, partly to lessen the cost of CAP to Britain. Secondly, for a time at least, the party had to accept membership of the EEC but remained committed to the reform of CAP. With the party's return to opposition, it was felt that not enough had been achieved and the 1980 party conference voted for Britain's withdrawal from the EEC. Once this position had been accepted Labour's agricultural policy returned to its traditional base of deficiency payments, but now combined with structural reform to shift Britain's pattern of agricultural production back towards livestock and horticulture (Labour Party 1981, 1983). Both these elements were to be allied with a new environmental awareness.

For the Conservatives policy development has been far more straight forward. The Conservative Party, and particularly its all important leadership, is still largely in favour of Britain remaining within the EEC and continues to accept the philosophy of CAP, though not its workings. As Michael Jopling made clear at the 1983 party conference improvements to CAP are generally held to be most fruitful if they improve the Community's price discipline to reduce surpluses and promote more efficient farming. Conservative

MEPs have generally supported this position (Jackson 1981, Pearce et al 1981).

CONCLUSION

There is obviously now a fundamental divide between the parties in the structure of their agricultural policies. This situation could actually be explained by a model of party competition. Labour's position would be perceived as an attempt to mollify its traditional supporters. Although it might be felt that the urban bias of the party continues to mean that the interests of the rural working class are marginalised. So this perspective is not without value and can help in understanding some of the dynamics of party policy-making. Moreover, the concept of party competition is widely utilised by politicians in presenting policies and ideas. Yet, on their own, models of party competition cannot plausibly explain why parties adopt the policies that they do. They are inadequate for a number of reasons.

First, such models ignore the organisational complexity of political parties and their means of policy-making. Party leaders, especially in the Labour Party, cannot unproblematically introduce electorally popular policies, even if they wished to. There can be severe limits on leaders' behaviour, as we saw with the Conservative Party and the mobilisation of its internal farming lobby in the early negotiations with the EEC. Further as the philosophy and dominant ideas within a party change over time there is a corresponding effect on policy, and such influences need not be related to party competition. Attitudes towards the EEC membership are a prime example.

Secondly, the role which interest groups perform in influencing the direction of party policy is ignored. The early post-war stategy of the NFU and other agricultural organisations was to remove agricultural policy from the sphere of party politics, and for a time it appeared that they might be successful. This is an important factor which cannot be ignored in accounting for the immediate post-war policies of the two major parties.

Finally, the thesis of party competition minimises the class based nature of politics. As we have seen agricultural policy-making was largely farmer orientated, and this bias has rarely been questioned. Even the Labour Party failed to overcome this bias in policy-making. The nearest it came was in its concern for the urban consumer. Yet even cheap food prices were not to be at the expense of farmers' welfare. It was a very different picture for the rural working class as represented by the farmworker. Generally their interests were overlooked by the Labour Party. Farmworkers are still amongst the poorest of the employed population, despite the profits which farmers have made. It was not until 1976 that the tied cottage was effectively curtailed despite numerous promises of abolition by the party since well before World War II. The Labour Party has proved unwilling to make up in political power what the farmworkers' union lacks in industrial muscle. Bias such as this, though, cannot easily be incorporated into a model of party competition which has to allow for freedom, within bounds, in the making of policy. Yet this is not competition but discrimination. There has been continued mobilisation of bias in agricultural policy, consistently favouring farmers at the expense of the rural working class. Policy has remained within safe 'bounds', as demarcated by the major political parties, the agricultural interest groups and the Ministry of Agriculture. This has helped to ensure that the basis of policy has for much of the post-war period remained unquestioned. Undoubtedly a problem for models of party competition when considered on their own is that they have conservative implications as they focus attention on the movements of parties and electors. This helps us to understand how policy is made but detracts attention from the important issues of who makes policy and in whose interest it is made.

Thus we have seen that even in the period characterised as one of party competition, the 1940s and 1950s, this is only one factor amongst others in determining party policy. A far better picture of why policies take the form that they do emerges from a study of internal party policy-making in which account is taken of the role of interest groups, party ideologies, the biases inherent within policy areas, and the competitive nature of party politics.

FOOTNOTES

1 Michael Parkinson (1970) in his study of Labour Party
   policy-making in relation to secondary education has argued that
   'one of the most crucial and enduring functions of parties...is
   the role they play in aggregating demands and interests, more
   simply in developing public policies for governing the community
   in which they exist'.

2 The most important problems with Downs work for this analysis
   relate to (i) party agreement with the majority of electors;
   (ii) party reliability and responsibility; (iii) office seeking
   by politicians; (iv) leadership unity, and (v) the existence of
   predispositional and short term cue voting. These difficulties
   are discussed in Budge and Farlie (1977) p113-19.

3 Robertson, in fact, highlights two ideas: a policy-method space
   and a policy aims continuum. He argues that parties have to deal
   with policy, with methods and proposed solutions and that it is
   necessary to see this activity as taking place in a policy-method
   space. However, the majority of voters are not distributed in
   this space, because they are interested in the results of
   government activity. As such they must be regarded as being
   distributed in a second policy aims space (or continuum) in which
   their desires can be represented. Though they are different these
   two spaces have to have points of congruence; a certain policy
   will bring about an impact on society which will be more or less
   desired by the electorate. In this way any policy-method point in
   one space can be mapped onto the other. As it is the distribution
   of voters' preferences that determines electoral results, the
   vote maximising position, in policy-method space represents, or
   is a mapping of, the actual majority preference for a particular
   outcome in the policy aims space. The spatial concept which Budge
   and Farlie utilise is that of the party defined space. This
   concept is developed from a different tradition from Robertson's;
   that of voting behaviour, particularly the social group and party
   identification theories. The party-defined space has end points
   which are made up to some kind of 'ideal' party position, and
   within which individuals are arrayed in terms of their degree of
   identification with or propensities towards the ideal parties
   (i.e. which party to vote or not to vote for). The propensities
   towards a party consist of both predispositions and cues.
   Robertson provides a working rational choice representation of
   actual election processes in Britain. Budge and Farlie then
   synthesise this tradition of party competition with that of
   voting behaviour. Robertson's notion of rationality has been
   severely criticised by Laver (1978). Instead, as Laver suggests,
   Budge and Farlie have tried to develop an a-rational explanation
   of voting choice combined with a rational choice theory of party
   competition without using Robertson's weak definition of
   rationality: see Budge (1978).

4 See the resolutions to the 1939 conference which congratulate the
   Agricultural Campaign Committee on the progress made in rural
   areas, which is regarded as a necessary step to winning the next
   General Election. The Labour Party, Report of the 38th Annual

Conference, Southport, May 29 - June 2 1939, P315-20. In the post-war campaign the General Secretary of the party argued that the 'response of the Rural Constituencies justifies all the special efforts which have been put forth in recent years. It is of vital importance that these efforts should be sustained, strengthened and made still more effective by Campaigns that will keep and extend Labour's hold in rural areas'. Party Development Report, 26 September, 1945.

5 Land nationalisation was regarded as a vote loser by some of the party leadership, and the party's General Secretary, Morgan Phillips, claimed that continued commitment to it meant that 'Tory opponents were able in many areas to foster irrational fears of land nationalisation'. Letter from Morgan Phillips to all members of the NEC, 15 March, 1950. Such concern remained, despite the issue being continually deprioritised and watered down during the period of Labour Government.

6 Butler was to write in a letter to Hurd: 'As I see it our main political task at the next Election will be to maintain farmers' confidence in the light of the suspension of the Brussels negotiations and the need to make changes in our method of support...we should do nothing which may appear to undermine the two Agriculture Acts (which) are fortunately very flexible'. The Parliamentary Agriculture Committee also warned of the dangers of losing rural constituencies if the party failed to support the 1947 and 1957 Agriculture Acts.

7 This was apparent in a letter from the Chairman, Sir Anthony Hurd to R.A. Butler, in February 1963. Copies of the letter were circulated to the Minister of Agriculture, Secretary of State for Scotland and also at a meeting of the Advisory Committee on Policy. A reply by R.A. Butler to Sir Anthony Hurd on 1 March, 1963, confirmed the position. Hallet and Jones (1963) believed that a minimum import prices policy, which the Government did implement, would stop importers from buying too cheaply, and was preferable to a quota system. Moreover 'politically this is an attractive plan, particularly to a government which is shortly to fight an election. By adhering to the principle of guaranteed prices, as outlined in the 1947 and 1957 Agriculture Acts, it can claim to have kept its faith with the farming community, as well as safeguarding the 'farming vote'.'

REFERENCES

Budge, I. 1978. Voter rationality and party competition: a comment on Laver's suggested approach. British Journal of Political Science 8, 511.
Budge, I. and Farlie, D. 1977. Voting and party competition. London: John Wiley.
Conservative Party, 1943. Sub-committee on agricultural reconstruction. London.
Conservative Party, 1947. Memorandum on second reading of the Agricultural Bill. London: Parliamentary Committee for Agriculture and Food.

Conservative Party, 1954. Parliamentary agriculture, fisheries and
    food committee. London.
Conservative Party, 1955. 75th Annual conference, Bournemouth.
    London.
Conservative Party, 1957a. 77th Annual conference, Brighton. London.
Conservative Party, 1957b. Parliamentary agriculture, fisheries and
    food committee. London.
Conservative Party, 1960. Farming questions answered. London:
    Conservative Central Office.
Conservative Party, 1966. Agriculture. London: Notes on current
    politics No 9.
Downs, A. 1957. An economic theory of democracy. New York: Harper
    and Row.
Hoffman, J. D. 1964. The Conservative Party in opposition, 1945-51.
    London: Macgibbon and Kee.
Gardner, T. W. 1979. Agricultural policy: formative influences in
    Britain. University of Manchester: Department of Agricultural
    Economics Bulletin 169.
Godber, J. 1970. The farming future. London: Conservative Political
    Centre.
Hallet, G. and Jones, G. 1963. Farming for consumers. London:
    Institute of Economic Affairs.
Jackson, R. 1981. From boom to bust? London: Conservative Political
    Centre.
Labour Party, 1946a. Report of the 45th annual conference,
    Bournemouth. London.
Labour Party, 1946b. Policy committee. London.
Labour Party, 1948. Report to policy committee of sub-committee on
    agriculture and rural life. London.
Labour Party, 1950. Draft policy statement. London: NEC.
Labour Party, 1956. Report of the 55th annual conference, Blackpool.
    London.
Labour Party, 1961. Britain and Europe: a draft report. London:
    Finance and economic policy sub-committee.
Labour Party, 1963a. Draft agricultural policy. London:
    Sub-committee on agriculture.
Labour Party, 1963b. Report of the 62nd annual conference,
    Scarborough. London.
Labour Party, 1970. Britain is strong - lets make it a great place
    to live. London: NEC Draft Manifesto.
Labour Party, 1971. Agriculture sub-committee. London: House of
    Commons.
Labour Party, 1975. The Common Market renegotiations: an appraisal.
    London: Background paper for special conference.
Labour Party, 1981. Withdrawal from the EEC. London: Research
    Department.
Labour Party, 1983. Agriculture - the Tory record. London: Research
    Department.
Laver, M. 1978. On defining voter rationality and deducing a model
    of party competition. British Journal of Political Science 8,
    253-6.
NFU, 1943. Agriculture and the nation. Interim report on post-war
    food production policy. London.
Parkinson, M. 1970. The Labour Party and the organisation of
    secondary education 1918-65. London: Routledge and Kegan Paul.

Pearce, A. et al. 1981. Food for Europeans. London: European
    Democratic Group.
RASE, 1941. Recommendations of a special committee appointed by the
    executive committee to consider post-war agricultural policy.
    London.
Robins, L. J. 1979. The reluctant party: Labour and the EEC,
    1961-75. Ormskirk: Hesketh.
Robertson, D. 1976. A theory of party competition. London: John
    Wiley.
Self, P. and Storing, H. 1962. The state and the farmer. London:
    George Allen and Unwin.

List of contributors

GRAHAM COX lectures in Sociology at the University of Bath. He has written a number of papers on agricultural and land-use politics and is co-author of Countryside Conflicts: The Politics of Farming, Forestry and Conservation, published by Gower. He is currently researching accountability and regulation in the dairy sector.

HARRIET FRIEDMANN is Associate Professor of Sociology at the University of Toronto. She has written widely on theories of the peasantry and family farming, and is currently completing a book entitled The Political Economy of Food.

ANDREW FLYNN is Research Assistant in the School of Planning at University College London, where he has recently completed a review of alternative uses for British agricultural land. He is finishing a Ph.D. thesis on rural politics and policy making.

RUTH GASSON is Senior Research Associate at Wye College. She has written numerous monographs and papers in agricultural economics and rural sociology. Her research has focused on social change in British agriculture, most recently concerning part-time farming and the role of women in agriculture.

DAVID GOODMAN lectures on Latin American Economics in the Department of Political Economy at University College London and in the Institute of Latin American Studies. He has written widely on the political economy of agriculture. He is co-author of From Peasant to Proletarian published by Basil Blackwell, who are also to publish his most recent study From Farming to Biotechnology.

PHILIP LOWE lectures in Countryside Planning at University College London. He has written widely on the environmental movement and countryside issues. He is co-author of Environmental Groups in Politics, published by Allen and Unwin and of Countryside Conflicts: The Politics of Farming, Forestry and Conservation.

TERRY MARSDEN is Senior Lecturer in Town Planning at South Bank
Polytechnic. He has written a number of papers on aspects of
capitalist agriculture in Britain. He is currently directing an ESRC
funded project on agricultural change in contemporary Britain.

JOAN MOSS is Senior Agricultural Economist in the Northern Ireland
Department of Agriculture and lectures at the Queen's University of
Belfast. As well as her interest in the agrarian economy of Northern
Ireland she has written on agriculture in the developing countries,
and spent two years as a Research Associate at the Institute of
Development Studies, Nairobi.

GEORGE PETERS is Research Professor in Agricultural Economics at the
University of Oxford. Prior to moving to Oxford in 1980 he was
Professor of Economic Science at the University of Liverpool. He has
written widely in agricultural economics principally on issues of
policy and land use.

CLIVE POTTER is Lecturer in the Economics of Agricultural and
Environmental Management at Wye College. He is author of Investing
in Rural Harmony published by the World Wildlife Fund. His current
research is on set-aside policies for UK agriculture.

MICHAEL REDCLIFT is Lecturer in Latin American Economics at Wye
College and at the Institute of Latin American Studies. He is author
of Development and the Environmental Crisis, published by Methuen,
and co-author of From Peasant to Proletarian published by Basil
Blackwell. He has written many papers on the sociology of
agricultural development in Latin America.

SARAH WHATMORE is Research Associate in Geography at University
College London, where she is working on a study of agricultural
change in Britain and the roles and relations of women in farming.

MICHAEL WINTER is Research Officer at the University of Bath. He has
written a number of papers on the sociology of agriculture and
land-use politics. He is co-author of Countryside Conflicts: The
Politics of Farming, Forestry and Conservation, and is currently
researching the politics of milk quotas.

Index